E

Information analysis of vegetation data

Tasks for vegetation science 10

Series Editors

HELMUT LIETH   HAROLD A. MOONEY

University of Osnabrück, F.R.G.   *Stanford University, Stanford,Calif., U.S.A.*

Information analysis of vegetation data

E. FEOLI, M. LAGONEGRO and L. ORLÓCI

University of Trieste, Italy
and
University of Western Ontario, Canada

1984 **DR W. JUNK PUBLISHERS**
a member of the KLUWER ACADEMIC PUBLISHERS GROUP
THE HAGUE / BOSTON / LANCASTER

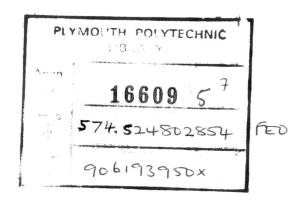

Distributors

for the United States and Canada: Kluwer Boston, Inc., 190 Old Derby Street, Hingham, MA 02043, USA
for all other countries: Kluwer Academic Publishers Group, Distribution Center, P.O.Box 322, 3300 AH Dordrecht, The Netherlands

Library of Congress Cataloging in Publication Data

```
Feoli, E.
   Information analysis of vegetation data.

   (Tasks for vegetation science ; 10)
   Includes bibliographical references.
   1. Plant communities--Statistical methods--Data
processing.  2. Vegetation classification--Statistical
methods--Data processing.  3. Botany--Ecology--
Statistical methods--Data processing.  4. Algorithms.
I. Lagonegro, M.  II. Orlóci, Laszló, 1932-
III. Title.  IV. Series.
QK911.F47 1984       581.5'248       83-19993
ISBN 90-6193-950-X
```

ISBN 90 6193 950 X (this volume)
ISBN 90 6193 897 X (series)

Cover design: Max Velthuijs

Copyright

PRINTED IN THE NETHERLANDS

Contents

Preface

Information analysis, a popular subject among vegetation ecologists not too many years ago, is revisited in this short monograph. The overview provided and the systematic presentation of ideas and algorithms should interest data analysts with backgrounds in this or other fields of natural science where the question of classification is addressed. The text gives the detailed descriptions and the listings of the computer programs.

The authors were recipients of grant support from the Italian Consiglio Nazionale delle Ricerche "Gruppo Biologia Naturalistica" (E. Feoli) and the Canadian National Science and Engineering Research Council (L. Orlóci) during completion of the project. The respective institutions of the University of Western Ontario and the University of Trieste provided facilities and computer time. Mrs. Stefani Tichbourne (London) typed the manuscript, Mr. Aulo Zampar (Trieste) gave computing assistance and Mr. Furio Poropat (Trieste) translated some programs. We are most grateful to them.

E. Feoli
M. Lagonegro
L. Orlóci

List of elementary symbols

E – event or equivocation

$I(E)$ – information content of event E

H – average information in a set of events

$P(E)$ – probability of event E

H^α – entropy of order α

P – observed distribution (relative frequencies) or a probability

P^0 – theoretical (expected) distribution or relative frequency

F – observed frequency distribution or a frequency

F^0 – theoretical (expected) frequency distribution or frequency

n – sample size

s – number of frequency classes (states)

χ^2 – chi-squared variate

α – order of entropy or a specified probability

v – degrees of freedom

J – J-divergence of two distributions

I – I-divergence of two distributions or a multiple of H

$h, i, j,$
u, z etc.   – subscripts identifying positions of elements in arrays

$h.$ – a row total; summation over second subscript

$.i$ – a column total; summation over first subscript

$..$ – grand total

N – number of individuals

X – occupancy counts

A – array F, N or X

$\overline{A}$ – average value

A^0 – theoretical or expected value

d – distance or relative divergence

r – coherence

X

,	–	separator symbol or expression of joint entropy, information
;	–	separator symbol or expression of mutual entropy, information
*	–	in formulae specifying label of an array in conditional state or removal of a subscript
Δ	–	divergence
t	–	number of arrays
Y	–	number of arrays
		–	a condition of something given indicated
B	–	number of blocks
b	–	block size
R	–	proportion of information accounted for

Information analysis of vegetation data

1. Introduction

The potential of information theory in Vegetation Science is broad and it entails possible applications in modelling, along the examples set by the early workers in cell biology (e.g. Yockey, Platzman & Quastler 1958), and quantification of statistical relationships as in the analytical methods of Kullback (1959). But the actual examples suggest to us that the role to be played in modelling is not as clearly understood as in data analysis.

Although in its broad generality data analysis poses similar problems in all fields, there are peculiar conditions existing which render the analysis of phytosociological and other vegetation data unique. These conditions are due to the fact that plant communities are aggregates of stationary objects, normally very complex and stratified, and often affected by seasonality. Also, the number of variables is usually very large. One major consequence of these is that the sampling units (plots, quadrats, stands) have to be arbitrarily delineated in space as well as in time. This influences strategy, and in turn, the stability of conclusions.

Many believe that for an understanding of community structure and function, precise quantitative description of *diversity* is required. This may mean the description of different properties related in some defined ways to *evenness* (the manner in which mass or numbers are arranged) and *multiplicity* (the richness factor). It is also believed useful to describe spatial relationships, such as in pattern analysis, classification, identification, or predictive analysis. The descriptions may however be based on different measures. But the measures most useful in complex problems are those which are additive, have scale independence, and known distribution. The sum of squares and the logarithmic expressions of proportions, such as entropy and information as defined in the next sections are of this type.

So much work has been done already on both theoretical and practical aspects in our field that a comprehensive review treating the theory and applications of entropy and information measures appears to be most useful at this time. In the following sections we offer such a review with respect to cluster analysis, identification, and prediction. We make passing references to the topics of diversity and spatial pattern.

2. Definitions of entropy and divergence

For simplicity, we shall divide the so-called information measures into entropy measures and divergence measures. Not all of these use logarithmic expressions. Fisher (1948), for example, defined information independently of logarithms. We shall, however, consider only the logarithmic expressions which we prefer, because they not only facilitate an additive analysis, but also they can be applied to nominal variables.

If event E has a probability $P(E)$ then the *information* associated with it may be defined as the quantity

$$I(E) = - \ln P(E). \tag{1}$$

The function $- \ln P(E)$ is always positive and its graph is monotone, descending to zero (Fig. 1). Since it increases with decreasing $P(E)$, it is sometimes described as a measure of the 'surprisal value' of E. It should be noted that for s independent events $E = 1, 2, \ldots, s$ from continuous distributions, the information

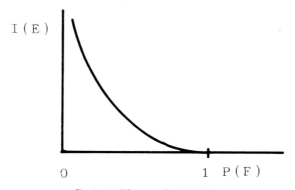

Figure 1. The graph of $I(E)$.

$$I(E) = - 2 \sum_{E=1}^{s} \ln \Pi(E) \tag{2}$$

is a chi square variate with $2s$ degrees of freedom (Fisher 1963). The $\Pi(E)$ is a tail probability.

If (1) is called information, $- P(E) \ln P(E)$ *is* E's contribution to the average information. This average is called *entropy* and it is defined by

$$H = - \sum_{E=1}^{s} P(E) \ln P(E) \tag{3}$$

The $P(E)$ in (3) are such that $\sum P(E) = 1$. Such a constraint does not apply to $\Pi(E)$ in (2). H of (3) has been variously described as Shannon's entropy function (Shannon 1948) or Rényi's entropy of order one (Rényi 1961). In fact, H is a special case of Rényi's (1961) measure for generalized entropy

$$H^\alpha = \frac{\ln \sum_{E=1}^{s} P(E)^\alpha}{1-\alpha} \tag{4}$$

when α approaches one.

Compared to Fig. 1, the graph of $- P(E) \ln P(E)$ is non-monotone (Fig. 2). H is a sum of ordinates if $P(E)$ is discrete, and an integral if continous. If normal, the entropy of the distribution becomes $\sqrt{2\pi e} \ln \sigma^2$ where σ^2 is the second moment.

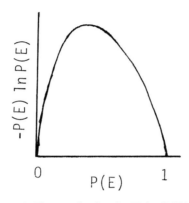

Figure 2. The graph of $- P(E) \ln P(E)$.

H^α has well known properties (Fadeev 1957, Kinchin 1957, Feinstein 1958, Basharin 1959, Rényi 1961). Noting that H is maximal when all $P(E)$ are equal, and minimum when all but one of the events occur only once, it has been suggested to call H a measure of *disorder*. It must be noted that order means the opposite of an equidistribution of frequencies and it is in no way indicative of the order relations in magnitude or occurrence of the events.

Whereas H measures disorder, the related information quantity of

$$H(P;P^0) = H(P^0) - H(P) = \sum_{E=1}^{s} P(E) \ln \frac{P(E)}{P^0(E)} \tag{5}$$

measures *divergence*. An observed s-valued distribution

$$P = [P(1)\ P(2) \dots P(s)]$$

is compared to another s-valued distribution

$$P^0 = [P^0(1)\ P^0(2) \dots P^0(s)]$$

called standard. P and P^0 are always equal valued, equitotaled, and identically ordered. P^0 is usually specified by hypothesis.

A multiple of $H(P;P^0)$,

$$I(F;F^0) = \sum_{E=1}^{s} F(E) \ln \frac{P(E)}{P^0(E)} \tag{6}$$

is called *information of order one* (Rényi 1961). In this, F and F^0 signify frequency distributions,

$$F = nP$$

$$F^0 = nP^0$$

n is sample size. It is noted that $s \leqslant n$, i.e. the number of observed states of a variable cannot be larger than the sample size. The quantity

$$\chi^2 = 2I(F;F^0) \tag{7}$$

is the *minimum discrimination information statistic* (Kullback 1959).

$I(F;F^0)$ is a non-symmetric measure of divergence. This means that the divergence $F^0 \to F$ is not the same as $F \to F^0$. $I(F;F^0)$ is called *I*-divergence information to distinguish it from the *J*-divergence information

$$J(F;F^0) = \sum_{E=1}^{s} [F(E) - F^0(E)] \ln \frac{P(E)}{P^0(E)} \tag{8}$$

which is symmetric.

A sampling distribution has been derived for $2I(F;F^0)$. It is indeed assymptotically a chi square distribution with given degrees of freedom (v) (Kullback 1959). Because of the approximate relationship of $2I(F;F^0)$ and χ^2,

$$2I(F;F^0) \simeq \sum_{E=1}^{s} \frac{[F(E) - F^0(E)]^2}{F^0(E)} = \chi^2 \tag{9}$$

and since the chi square distribution may only be an approximation to the true sampling distribution of χ^2, the procedure which finds probabilities for $2I(F;F^0)$ based on the chi square distribution, i.e.,

$$P(\chi^2 \geqslant \chi^2_{\alpha;v}) = \alpha \tag{10}$$

has to be regarded as a potentially weak approximation.

Dispersion, diversity, or predictive arrays are constructed to facilitate the computations. These are discussed in the following sections.

3. Arrays of vegetation data

R-dispersion arrays

These arrays describe the variables singly or jointly. They contain elements which are counts of discrete events. An event is the occurrence of a given state in a given variable. For example, for a variable h of s_h discrete states, a dispersion array is constructed by counting the number of occurrences of each state in $n \geqslant s_h$ observed units. Assume that the jth state occurs F_{hj} times. Then the dispersion array

$$F_h = [F_{h1} \ F_{h2} \ \ldots \ F_{hsh.}] \tag{11}$$

is the frequency distribution of variable h. Since this does not take into account joint occurrences with other variables, F_h is called a *marginal dispersion array* or *marginal distribution*. Should there be more than two variables in the set, F_h could be called a *principal marginal distribution*.

Variables are often considered jointly in twos, threes, etc. If such is the case, not only the marginal distributions are of interest but also the p-dimensional dispersion arrays which describe their joint distributions. For two variables h and i, the joint distribution is two dimensional,

$$F_{hi} = \begin{bmatrix} F_{h1,i1} & F_{h1,i2} & \cdots & F_{h1,is_i} \\ F_{h2,i1} & F_{h2,i2} & \cdots & F_{h2,is_i} \\ \cdot & \cdot & \cdots & \cdot \\ F_{hsh,i1} & F_{hsh,i2} & \cdots & F_{hsh,is_i} \end{bmatrix}.$$

An element in the intersection of the jth row (state of variable h) and kth column (state of variable i) is $F_{hj,ik}$. This is the number of joint occurrences of the two states in the n observed units. The marginal totals

$$F_{hj,i.} = \sum_{k=1}^{si} F_{hj,ik} = F_{hj}$$

$$F_{h.,ik} = \sum_{j=1}^{sh} F_{hj,ik} = F_{ik} \tag{12}$$

are elements in F_h or F_i as in (11). The grand total is

$$F_{h.,i.} = \sum_{j=1}^{sh} \sum_{k=1}^{si} F_{hj,ik} = F_{..} \tag{13}$$

Since the symbolism used may not be familiar, an example should clarify the intended meaning (Table 1).

6

Table 1. Construction of an *R*-dispersion array for a two-species set. The elements in the raw data represent occupancy counts. The elements in the dispersion arrays represent frequencies. The states are arbitrary.

Raw data

	Relevés
Arrhenatherum elatius	3 2 0 4 0 0 0 0 20 0 20 0 2 0 7 2 0 0 0 0 5 0 0 20 8 12 0 25 22 0 0 2 2 0 0 0 0
Bromus erectus	20 10 5 0 0 0 17 0 0 8 0 8 18 8 4 0 18 3 0 5 4 2 15 6 5 20 10 15 15 9 10 4 6 25 25 15 24

Marginal dispersion arrays

Class limits	0	1–9	10–15	16–25
Arrhenatherum elatius	20	10	1	4
Bromus erectus	7	12	6	10

R-dispersion array

		Bromus erectus				Total
		0	1–9	10–15	16–25	
Arrhenatherum elatius	0	4	7	3	6	20
	1–9	2	4	1	3	10
	10–15	0	0	0	1	1
	16–25	1	1	2	0	4
	Total	7	12	6	10	35

Q-dispersion arrays

These describe the individuals singly or jointly. Assume that in individuals j and k the p variables have s_j and s_k states. The joint occurrences of states are counted, irrespective of variables, and summarized in a $s_j \times s_k$ *dispersion array*:

$$N_{jk} = \begin{bmatrix} N_{j1,k1} & N_{j1,k2} & \cdots & N_{j1,ks_k} \\ N_{j2,k1} & N_{j2,k2} & \cdots & N_{j2,ks_k} \\ \cdot & \cdot \cdot \cdot & \cdot \cdot \cdot \cdot & \cdot \cdot \cdot \cdot \\ N_{js_j,k1} & N_{js_j,k2} & \cdots & N_{js_j,ks_k} \end{bmatrix}.$$

The $N_{jl,km}$ are not comparable to the $F_{hj,ik}$ as frequencies. The variables are not distinguished, only the individuals are isolated. The marginal totals

$$N_{jl,k.} = \sum_{m=1}^{sk} N_{jl,km} = N_{jl}$$

$$N_{j.,km} = \sum_{l=1}^{sj} N_{jl,km} = N_{km}$$

(14)

indicate either the number of variables with state l in individual j or with state m in individual k. The grand total is $N_{j.,k.}$; this is equal to p, the number of variables. Table 2 contains an example.

Diversity arrays

When the observations are occupancy counts or counts of individuals, such as in density data, the data matrix

$$X = \begin{bmatrix} X_{11} & X_{12} & \cdots & X_{1n} \\ X_{21} & X_{22} & \cdots & X_{2n} \\ \cdot & \cdot & \cdots & \cdot \\ X_{p1} & X_{p2} & \cdots & X_{pn} \end{bmatrix}$$

is called a *diversity array*. The rows correspond to variables and the columns to individuals. A general element X_{ij} is the value of variable i in individual j. X is called a diversity array, since it conveys information about evenness and richness. The totals

8

8

Table 2. Construction of a Q-dispersion array for two reléves based on 34 species. The elements in the raw data are occupancy counts. The elements in the dispersion arrays are counts of variables in given states. The states are arbitrary.

Raw data

Species

| Relevé 1 | 0 | 7 | 1 | 5 | 6 | 0 | 0 | 0 | 4 | 10 | 22 | 20 | 4 | 20 | 5 | 0 | 20 | 2 | 0 | 0 | 0 | 0 | 1 | 0 | 0 | 3 | 0 | 7 | 3 | 25 | 4 | 20 | 0 |
| Relevé 2 | 4 | 0 | 1 | 4 | 5 | 6 | 0 | 6 | 0 | 0 | 0 | 0 | 6 | 5 | 5 | 5 | 6 | 7 | 7 | 0 | 0 | 0 | 7 | 5 | 0 | 0 | 3 | 3 | 2 | 10 | 3 | 10 | 13 |

Marginal dispersion arrays

Class limits	0	1-9	10-15	16-25
Relevé 1	14	12	2	6
Relevé 2	13	18	3	0

Q-dispersion array

		Relevé 2			
		0	1-9	10-15	Total
Relevé 1	0	7	6	1	14
	1-9	3	9	0	12
	10-15	1	1	0	2
	16-25	2	2	2	6
	Total	13	18	3	34

$$X_{i.} = \sum_{j=1}^{n} X_{ij}$$

(15)

$$X_{.j} = \sum_{i=1}^{p} X_{ij}$$

define marginal distributions. The grand total is

$$X_{..} = \sum_{i=1}^{p} \sum_{j=1}^{n} X_{ij} .$$

(16)

Predictive arrays

The notion of joint occurrence entails the idea of prediction. The prediction is symmetrical and its strength is dependent on the structure within F_{hi}, N_{jk} or X. Should a clear trend exist, the prediction would be strong. On this basis, every matrix of counts or frequencies is by definition a predictive array. F_{hi} and N_{jk} are not related to order in the measuring scale of the states, therefore they do not contain information about order-related interactions. The implication of this is that if F_{hi} or N_{jk} are analysed, the information residing in the magnitude of the measurements will be lost. X is exceptional in that its elements are counts and thus in them the order relations in the measuring scale continue to exist. Arrays F_{hi}, N_{jk} and X are indeed carriers of information about association or resemblance, which are functions of symmetric prediction.

Notes on symbols

Since the information measures to be applied are similar in form irrespective of F, N and X, these symbols will be replaced by the equivalent symbol A. If the formulations refer to F or N, the involvement of a pair (u,z) is meant. In general, u,z will represent variables if F is meant, or individuals if N is meant. If X is given, u will represent variables and z will individuals. When the states in u or z are meant the subscripts j and k will be used. The number of variables will be designated by p and the number of individuals by n. The symbols s_u or s_z signify numbers of states. In the case of array X, the formulae will contain no subscripts for j,k but only u,z. When the symbol $\overline{A}$ is used, a mean value or an equidistribution is meant. Symbol A^0 will mean random expectation according to hypothesis. In the symbols a superscript within parentheses is a label for the array.

4. Measurements on the arrays

Entropy measures and multiples

Measures on the marginal totals can be defined as multiples of (3). These will decompose diversity if A_{uz} is interpretable as a diversity matrix:

$$I(A_u) = -\sum_{j=1}^{su} A_{uj,z.} \ln A_{uj,z.}/A_{u.,z.}$$

$$= A_{u.,z.} \ln A_{u.,z.} - \sum_{j=1}^{su} A_{uj,z.} \ln A_{uj,z.} \qquad (17)$$

$$I(A_z) = -\sum_{k=1}^{sz} A_{u.,zk} \ln A_{u.,zk}/A_{u.,z.}$$

$$= A_{u.,z.} \ln A_{u.,z.} - \sum_{k=1}^{sz} A_{u.,zk} \ln A_{u.,zk} \qquad (18)$$

A measure can also be defined on the elements in A_{uz} as a multiple of (3),

$$I(A_{uz}) = I(A_u, A_z) = -\sum_{j=1}^{su}\sum_{k=1}^{sz} A_{uj,zk} \ln A_{uj,zk}/A_{u.,z.}$$

$$= A_{u.,z.} \ln A_{u.,z.} - \sum_{j=1}^{su}\sum_{k=1}^{sz} A_{uj,zk} \ln A_{uj,zk} \qquad (19)$$

A multiple of (5) is (17) plus (18) minus (19):

$$I(A_{uz};A_{uz}^0) = I(A_u;A_z) = I(A_u) + I(A_z) - I(A_u,A_z)$$

$$= \sum_{j=1}^{su}\sum_{k=1}^{sz} A_{uj,zk} \ln \frac{A_{uj,zk} A_{u.,z.}}{A_{uj,z.} A_{u.,zk}}. \qquad (20)$$

The fraction

$$\frac{A_{uj,z.} A_{u.,zk}}{A_{u.,z.}}$$

is the jk element in A^0, normally called the *expectation array*.

These different quantities can be represented by a Venn diagram (Fig. 3). The area of overlap is proportional to the divergence of A_{uz} from A_{uz}^0. This is the information held in common between A_u and A_z, the *mutual information* which measures predictive value:

$I(A_u,A_z)$ is called *joint information*, and the symmetric difference

$$E(A_u;A_z) = I(A_u,A_z) - I(A_u;A_z) \qquad (21)$$

is the *equivocation information*.

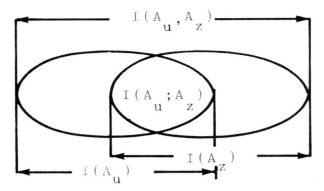

Figure 3. Venn diagram.

The quantities (17) to (21) are not directly comparable between samples, since their magnitude is dependent on sample size. The relative measures do not have this difficulty. To relativize (17), divide $I(A_u)$ by $A_{u.,z.}$ ln s_u. A similar division will relativize (18). Division by $A_{u.,z.}$ ln $s_u s_z$ is required for (19) and by the smallest of (17) or (18) for (20).

Relative measures are broadly used. One is Rajski's (1961) metric (22) and the other the coherence coefficient (23):

$$d(A_u;A_z) = \frac{E(A_u;A_z)}{I(A_u,A_z)} \tag{22}$$

$$r(A_u;A_z) = \sqrt{1 - d^2\,(A_u;A_z)}\ . \tag{23}$$

Whereas the Rajski metric measures the difficulty in predicting A_u based on A_z, or A_z based on A_u, the coherence coefficient measures directly the strength of the prediction. All formulations (17) to (23) can be extended to apply to any p variables simultaneously. An example is given in Table 3.

Table 3. The following example illustrates the calculations. The raw data is the Q-dispersion array in Table 2. The entropy quantities accord with formulae (17) to (25).

$I(A_u)$	$=$	$-(14 \ln \frac{14}{34} + 12 \ln \frac{12}{34} + 2 \ln \frac{2}{34} + 6 \ln \frac{6}{34})$
	$=$	$34 \ln 34 - 14 \ln 14 - 12 \ln 12 - 2 \ln 2 - 6 \ln 6$
	$=$	40.99
$I(A_z)$	$=$	$-(13 \ln \frac{13}{34} + 18 \ln \frac{18}{34} + 3 \ln \frac{3}{34})$
	$=$	$34 \ln 34 - 13 \ln 13 - 18 \ln 18 - 3 \ln 3$
	$=$	31.23

Table 3 (continued).

$$I(A_u, A_z) = -(7 \ln \frac{7}{34} + 6 \ln \frac{6}{34} + \ln \frac{1}{34} + 3 \ln \frac{3}{34} + 9 \ln \frac{9}{34} + \ln \frac{1}{34} +$$
$$\ln \frac{1}{34} + 2 \ln \frac{2}{34} + 2 \ln \frac{2}{34} + 2 \ln \frac{2}{34})$$
$$= 34 \ln 34 - 7 \ln 7 - 6 \ln 6 - 3 \ln 3 - 9 \ln 9 - 2 \ln 2 - 2 \ln 2 - 2 \ln 2$$
$$= 68.28$$

$$I(A_u : A_z) = 40.99 + 31.23 - 68.28$$
$$= 3.94$$

$$E(A_u : A_z) = 68.28 - 3.94$$
$$= 64.34$$

$$d(A_u, A_z) = 0.94$$

$$r(A_u : A_z) = \sqrt{1 - 0.88}$$
$$= \sqrt{0.12}$$
$$= 0.35$$

$$I(A_{u \cdot}) = -(7 \ln \frac{7}{14} + 6 \ln \frac{6}{14} + \ln \frac{1}{14} + 3 \ln \frac{3}{12} + 9 \ln \frac{9}{12} + \ln \frac{1}{2} + \ln \frac{1}{2} +$$
$$2 \ln \frac{2}{6} + 2 \ln \frac{2}{6} + 2 \ln \frac{2}{6})$$
$$= 14 \ln 14 - 7 \ln 7 - 6 \ln 6 + 12 \ln 12 - 3 \ln 3 - 9 \ln 9 + 2 \ln 2 +$$
$$6 \ln 6 - 2 \ln 2 - 2 \ln 2 - 2 \ln 2$$
$$= 27.30$$

$$I(A_{z \cdot}) = -(7 \ln \frac{7}{13} + 3 \ln \frac{3}{13} + \ln \frac{1}{13} + 2 \ln \frac{2}{13} + 6 \ln \frac{6}{18} + 9 \ln \frac{9}{18} +$$
$$\ln \frac{1}{18} + 2 \ln \frac{2}{18} + \ln \frac{1}{3} + 2 \ln \frac{2}{3})$$
$$= 13 \ln 13 - 7 \ln 7 - 3 \ln 3 - 2 \ln 2 + 18 \ln 18 - 6 \ln 6 - 9 \ln 9 -$$
$$2 \ln 2 + 3 \ln 3 - 2 \ln 2$$
$$= 37.05$$

Components of entropy

A. A single array

For a given A_{uz} the components recognized are described in the following table:

Source	Multiple of entropy	Formula
Between rows	$I(A_u)$	(17)
Between columns	$I(A_z)$	(18)
Interaction	$I(A_{uz};A_{uz}^0)$	(20)
Joint	$I(A_u,A_z)$	(19)

The components may be formulated for $I(A_u,A_z)$ based on the rows:

Source	Information	Formula
Between rows	$I(A_u)$	(17)
Within rows	$I(A_{u*}) = -\sum_{j=1}^{su} \sum_{k=1}^{sz} A_{uj,zk} \ln A_{uj,zk}/A_{uj,z.}$	(24)
Joint	$I(A_u,A_z) = I(A_u) + I(A_{u*})$	(19)

A partition may also be derived based on the columns:

Source	Diversity	Formula
Between columns	$I(A_z)$	(18)
Within columns	$I(A_{z*}) = -\sum_{j=1}^{su} \sum_{k=1}^{sz} A_{uj,zk} \ln A_{uj,zk}/A_{u.,zk}$	(25)
Joint	$I(A_u,A_z) = I(A_z) + I(A_{z*})$	(19)

Numerical examples are in Table 3.

B. Several arrays

It may be of interest to compare arrays whose dimensions are compatible in the direction of columns or in the direction of rows. When the comparison is in the direction of columns, the t arrays

$$A_{uz}^{(1)}$$
$$A_{uz}^{(2)}$$
$$A_{uz}^{(t)}$$
(26)

must have an equal number of columns. When the comparison is in the direction of rows, the v arrays

14

$$A_{uz}^{(1)}, A_{uz}^{(2)}, \ldots, A_{uz}^{(v)} \tag{27}$$

must have an equal number of rows.

Assuming two arrays $A_{uz}^{(1)}$ and $A_{uz}^{(2)}$, the following cases of entropy difference are commonly used (illustrated in Table 4).

Table 4. The calculations illustrated pertain to formulae of entropy differences or divergences of several arrays. The data of Table 1 are used, but arranged in two groups according to third dendrogram in Fig. D1.

The following are the data tables:

Array 1

	Relevés																			Totals
	1	22	11	19	13	21	27	29	4	9	2	25	23	24	26	30	17	35	28	19
Arrhenatherum elatius	3	20	2	5	7	2	22	0	4	20	2	0	8	12	25	2	0	0	0	134
Bromus erectus	20	6	18	4	4	6	15	10	0	0	10	10	5	20	15	4	0	24	9	180

Array 2

	Relevés																Totals
	3	34	7	10	12	21	15	32	33	5	6	14	8	20	16	18	16
Arrhenatherum elatius	0	0	0	0	0	0	0	0	0	0	0	2	0	0	0	0	2
Bromus erectus	5	15	17	8	8	15	18	25	0	0	0	0	0	2	3	5	146

All calculations are based on row totals –

134	2	136
180	146	326
314	148	462

Entropy difference by rows –

$$D_{uz}^{(1.2)1} = -[134 \ln 134/136 + 2 \ln 2/136 + 180 \ln 180/326 + 146 \ln 146/326]$$
$$= 234.613983$$

Entropy difference by interaction –

$$D_{uz}^{(1.2)3} = 134 \ln 134(462)/(136(314)) + 2 \ln 2(462)/(136(148)) +$$
$$180 \ln 180(462)/(326(314)) + 146 \ln 146(462)/(326(148))$$
$$= 55.1201813$$

Entropy difference irrespective of rows or columns –

$$D_{uz}^{(1.2)5} = -[314 \ln 314/462 + 148 \ln 148/462]$$
$$= 289.734166$$

Table 4 (continued).

Divergence by rows –

$$\Delta_{uz}^{(1,2)1} = 134 \ln 134/(19(136/35)) + 2 \ln 2/(16(136/35)) + 180 \ln 180/(19(326/35))$$
$$+ 146 \ln 146/(16(326/35))$$
$$= 73.0598508$$

Divergence by interaction –

$$\Delta_{uz}^{(1,2)2} = D_{uz}^{(1,2)3} = 55.1201813$$

Divergence by grand total –

$$\Delta_{uz}^{(1,2)3} = 314 \ln 314/(2(19)(462/(2(35)))) + 148 \ln 148/(2(16)(462/(2(35))))$$
$$= 17.3396686$$

The additive sequences in (33) and (41) can be verified.

---

(1) Entropy difference by rows

If the two arrays are joined by rows, the increase in entropy is

$$D_{uz}^{(1,2)1} = I(A_{u*})^{(1+2)} - I(A_{u*})^{(1)} - I(A_{u*})^{(2)}$$

$$= - \sum_{j=1}^{su} \left[\sum_{l=1}^{2} \sum_{k=1}^{s_z^{(1)}} A_{uj,zk}^{(l)} \ln \frac{A_{uj,zk}^{(l)}}{A_{uj,z.}^{(1)} + A_{uj,z.}^{(2)}} \right.$$
$$\left. - \sum_{l=1}^{2} \sum_{k=1}^{s_z^{(1)}} A_{uj,zk}^{(l)} \ln \frac{A_{uj,zk}^{(l)}}{A_{uj,z.}^{(l)}} \right]$$
$$= - \sum_{j=1}^{su} \sum_{l=1}^{2} A_{uj,z.}^{(l)} \ln \frac{A_{uj,z.}^{(l)}}{A_{uj,z.}^{(1)} + A_{uj,z.}^{(2)}} . \tag{28}$$

(2) Entropy difference by columns

If the two arrays are attached columnwise their entropy difference is equal to

$$D_{uz}^{(1,2)2} = I(A_{z*})^{(1+2)} - I(A_{z*})^{(1)} - I(A_{z*})^{(2)}$$

$$= - \sum_{l=1}^{2} \sum_{k=1}^{sz} A_{u.,zk}^{(l)} \ln \frac{A_{u.,zk}^{(l)}}{A_{u.,zk}^{(1)} + A_{u.,zk}^{(2)}} \tag{29}$$

The derivation is similar to (28).

16

(3) Entropy difference by interaction

If the union is by rows, the entropy difference is

$$D_{uz}^{(1,2)3} = I(A_{uz};A_{uz}^0)^{(1+2)} - I(A_{uz};A_{uz}^0)^{(1)} - I(A_{uz};A_{uz}^0)^{(2)}$$

$$= \sum_{j=1}^{su} \sum_{l=1}^{2} A_{uj,z}^{(l)} \cdot \ln \frac{A_{uj,z.}^{(l)} \cdot [A_{u.,z.}^{(1)} + A_{u.,z.}^{(2)}]}{[A_{uj,z.}^{(1)} + A_{uj,z.}^{(2)}] \, A_{u.,z.}^{(l)}} \tag{30}$$

For union by columns,

$$D_{uz}^{(1,2)4} = \sum_{l=1}^{2} \sum_{k=1}^{sz} A_{u.,zk}^{(l)} \ln \frac{A_{u.,zk}^{(l)} \cdot [A_{u.,z.}^{(1)} + A_{u.,z.}^{(2)}]}{A_{u.,z.}^{(l)} \cdot [A_{u.,zk}^{(1)} + A_{u.,zk}^{(2)}]} \tag{31}$$

(4) Entropy difference irrespective of rows or columns

If the row and column classifications are disregarded, i.e., the arrays $A_{uz}^{(1)}$ and $A_{uz}^{(2)}$ are regarded as two strings of numbers, the entropy difference is

$$D_{uz}^{(1,2)5} = I(A_{uz},A_{uz}^0)^{(1+2)} - I(A_{uz},A_{uz}^0)^{(1)} - I(A_{uz},A_{uz}^0)^{(2)}$$

$$= - \sum_{l=1}^{2} A_{u.,z.}^{(l)} \cdot \ln \frac{A_{u.,z.}^{(l)}}{A_{u.,z.}^{(1)} + A_{u.,z.}^{(2)}} \tag{32}$$

Of these, (28) and (29) give the potentially most restrictive comparison and (32) gives the least restrictive one. The following additive sequences are noted:

$$D_{uz}^{(1,2)5} = D_{uz}^{(1,2)1} + D_{uz}^{(1,2)3}$$

$$D_{uz}^{(1,2)5} = D_{uz}^{(1,2)2} + D_{uz}^{(1,2)4} \tag{33}$$

All formulae easily extend to any number of arrays.

Divergence measures

These can be formulated for a single array or for several of them simultaneously. Formulae (22), (30) and (31) are of this kind. They measure divergences from expected distributions. There are several types:

A. A single $s_u \times s_z$ array

(1) Main effect by rows

$$I(A_{u*,z.};\overline{A}_{u*,z.}) = \sum_{j=1}^{su} A_{uj,z.} \ln \frac{A_{uj,z.} \cdot s_u}{A_{u.,z.}} . \tag{34}$$

(2) Main effect by columns

$$I(A_{u.,z*};\overline{A_{u.,z*}}) = \sum_{k=1}^{s_z} A_{u.,zk} \ln \frac{A_{u.,zk}\, s_z}{A_{u.,z.}} \,. \tag{35}$$

(3) Interaction

$$I(A_{uz};A_{uz}^0) = \sum_{j=1}^{s_u} \sum_{k=1}^{s_z} A_{uj,zk} \ln \frac{A_{uj,zk}\, A_{u.,z.}}{A_{uj,z.}\, A_{u.,zk}} \,. \tag{36}$$

(4) Joint effect

$$I(A_{u*,z*};\overline{A_{u.,z.}}) = \sum_{j=1}^{s_u} \sum_{k=1}^{s_z} A_{uj,zk} \ln \frac{A_{uj,zk}\, s_u s_z}{A_{u.,z.}} \,. \tag{37}$$

The components are additive (joint effect representing the total). They associate with different degrees of freedom:

Source	Divergence	Formula	Degrees of freedom
Main effect by rows	$I(A_{u*,z.};\overline{A_{u*,z.}})$	(34)	$s_u - 1$
Main effect by columns	$I(A_{u.,z*};\overline{A_{u.,z.*}})$	(35)	$s_z - 1$
Interaction	$I(A_{uz};A_{uz}^0)$	(36)	$(s_u - 1)(s_z - 1)$
Joint effect	$I(A_{u*,z*};\overline{A_{u.,z.}})$	(37)	$s_u s_z - 1$

Similar partitions can be derived for more than two dimensional arrays but the numbers of source may increase quickly to the point of unmanageability.

B. Several arrays

The arrays considered have compatible dimensions in the direction of the comparison. For example, in the set

$$A_{uz}^{(1)}, A_{uz}^{(2)}, \ldots, A_{uz}^{(v)}$$

all the A_{uz} should have an equal number of rows, and in the set

$$A_{uz}^{(1)}$$
$$A_{uz}^{(2)}$$
$$.$$
$$A_{uz}^{(t)}$$

18

all should have an equal number of columns. Different information measures can be formulated on the divergence. If $t = 2$, a quantity Δ is defined measuring the information gain when two arrays $A_{uz}^{(1)}$ and $A_{uz}^{(2)}$ are joined. Three cases are common (see Table 4):

(1) Divergence by rows

$$\Delta_{uz}^{(1,2)1} = I(A_{uz};\overline{A}_{uz})^{(1+2)} - I(A_{uz};\overline{A}_{uz})^{(1)} - I(A_{uz};\overline{A}_{uz})^{(2)}$$

$$= \sum_{j=1}^{s_u} \sum_{l=1}^{2} A_{uj,z.}^{(l)} \cdot \ln \frac{A_{uj,z.}^{(l)}/n_{uz}^{(l)}}{(A_{uj,z.}^{(1)} + A_{uj,z.}^{(2)})/(n_{uz}^{(1)} + n_{uz}^{(2)})} . \tag{38}$$

This assumes that the row divergences are additive. $n_{uz}^{(1)}$ is the number of columns in $A_{uz}^{(1)}$ and $n_{uz}^{(2)}$ is the number of columns in $A_{uz}^{(2)}$. $\Delta_{uz}^{(1,2)}$ is a discrimination information measure with degrees of freedom s_u.

(2) Divergence by interaction

$$\Delta_{uz}^{(1,2)2} = I(A_{uz};A_{uz}^0)^{(1+2)} - I(A_{uz};A_{uz}^0)^{(1)} - I(A_{uz};A_{uz}^0)^{(2)} . \tag{39}$$

This is in fact the same as (30). (31) applies when the arrays are joined by the rows. Clearly, these measure the mutual information in an s_u by 2 or s_z by 2 array whose elements are the row totals of the dispersion arrays. Accordingly, the relevant degrees of freedom are $s_u - 1$ for (30, 39) and $s_z - 1$ for (31).

(3) Divergence by grand totals

The least rigorous measure of divergence involves the grand totals:

$$\Delta_{uz}^{(1,2)3} = \sum_{l=1}^{2} A_{u.,z.}^{(l)} \cdot \ln \frac{A_{u.,z.}^{(l)}/(s_u^{(l)} s_z^{(l)})}{[A_{u.,z.}^{(1)} + A_{u.,z.}^{(2)}]/[s_u^{(l)}(s_z^{(1)} + s_z^{(2)})]} \tag{40}$$

This is associated with a single degree of freedom.

The formulae provide for an additive analysis in the sense of

$$\Delta_{uz}^{(1,2)1} = \Delta_{uz}^{(1,2)2} + \Delta_{uz}^{(1,2)3} \tag{41}$$

and can be simply expanded to any numbers of dispersion matrices.

Information measures on diversity arrays

The measures formulated are applicable, but with a possibly different meaning. The reason is that whereas X is always a two dimensional array including p variables and n individuals simultaneously, F and N can be multidimensional. Furthermore in X, sets of different entities are related, such as variables in one set and individuals in another. This is not so in F and N which are specific to variables (F) or individuals (N).

Hierarchically nested model of divergence

In certain problems, hierarchical nesting of samples may occur. At each level, there are blocks and each block contains a given number of elements. When moving up in the hierarchy, the number of blocks is reduced and the number of elements increased within blocks. At the top there is always just a single block. At the bottom, the blocks have single elements. Considering the 4 lowest levels of an hierarchy, the following represents the symbolic data:

At the top level there are blocks,

$$[A_1 \ldots \quad A_2 \ldots \quad \ldots \quad A_j \ldots \quad \ldots \quad A_b \ldots] \; .$$

Each of these may have more than one second order blocks. For example, $A_j \ldots$ has b_j such blocks,

$$A_j \ldots = [A_{j1} \ldots \quad A_{j2} \ldots \quad \ldots \quad A_{jk} \ldots \quad \ldots \quad A_{jb_j}] \; .$$

By similar reasoning

$$A_{jk} \ldots = [A_{jk1} \ldots \quad A_{jk2} \ldots \quad \ldots \quad A_{jkl} \ldots \quad \ldots \quad A_{jkb_{jk}}]$$

and

$$A_{jkl} \ldots = [A_{jkl1} \quad A_{jkl2} \quad \ldots \quad A_{jklm} \quad \ldots \quad A_{jklb_{jkl}}] \; .$$

On the first level (bottom), A_{jklm} is the designation of a genaral block. Each block on this level contains only one element, so the block size symbol s_{jklm} always signifies unity. On the other levels of nesting, the number of elements vary among the blocks, e.g., s_j, s_{jk}, s_{jkl}. When there is more than one element per block the As will designate block totals and A/s the block means. Extension of the symbols is by addition of new subscripts at the bottom to accommodate more than four hierarchical levels. The components of information are given by formulae in Table 5 (Orlóci 1971). The vth difference measures the reduction in information divergence by collapsing the blocks of the vth level into the blocks of the $v + 1$st level. The collapsing of blocks follows a predetermined pattern. An example is given in Table 6.

Redundancy

Redundancy is a property of a set or sets of variables. It is conditionally defined in a variable or a set, assuming interaction with another variable or set. Redundancy is not usually symmetric and different functions may be used to measure it.

Table 5. Divergence measures in an hierarchically nested model. A single variable and four hierarchical levels are involved. $A_{....}$ and s are grand totals. Other symbols are explained in the main text.

Level in hierarchy	Divergence (I_v)	Information specific to level* Δ_v	Number of blocks B_v	Degrees of freedom
Bottom	$I_1 = \sum\limits_{j=1}^{b} \sum\limits_{k=1}^{b_j} \sum\limits_{l=1}^{b_{jk}} \sum\limits_{m=1}^{b_{jkl}} A_{jklm} \ln \dfrac{A_{jklm}/s_{jklm}}{A_{....}/s}$	$\Delta_1 = I_1 - I_2$	$B_1 = \sum\limits_{j=1}^{b} \sum\limits_{k=1}^{b_j} \sum\limits_{k=1}^{b_{jk}} \sum\limits_{l=1}^{b_{jk}} b_{jkl}$	$B_1 - B_2$
	$I_2 = \sum\limits_{j=1}^{b} \sum\limits_{k=1}^{b_j} \sum\limits_{l=1}^{b_{jk}} A_{jkl.} \ln \dfrac{A_{jkl.}/s_{jkl}}{A_{....}/s}$	$\Delta_2 = I_2 - I_3$	$B_2 = \sum\limits_{j=1}^{b} \sum\limits_{k=1}^{b_j} b_{jk}$	$B_2 - B_3$
	$I_3 = \sum\limits_{j=1}^{b} \sum\limits_{k=1}^{b_j} A_{jk..} \ln \dfrac{A_{jk..}/s_{jk}}{A_{....}/s}$	$\Delta_3 = I_3 - I_4$	$B_3 = \sum\limits_{j=1}^{b} b_j$	$B_3 - B_4$
Top	$I_4 = \sum\limits_{j=1}^{b} A_{j...} \ln \dfrac{A_{j...}/s_j}{A_{....}/s}$	$\Delta_4^{**} = I_4 - I_5$	$B_4 = b$	$B_4 - B_5^{***}$

* Divergence measured between v-level units within blocks of $v + 1$ level units.

** At level 4 and higher if higher levels exist.

*** If $B_5 = 1$ then $I_5 = 0$.

Table 6. Example of application of the nested model. A single species is involved. Only four levels are considered in a hierarchy of 5 levels before the complete fusion. The first level corresponds to data. The blocks are separated by asterisks.

Relevé	1	2	3	4	5	6	7	8	9	10	11	12	13	14	15	16	17	18	19	20	21	22	23	24	25	26	27	28	29	30	31	32	33	34	35
Bromus erectus	0	0	0	2	3	0	0	0	18	4	8	15	5	5	15	8	18	15	15	0	24	9	17	6	6	4	10	20	10	10	20	15	4	5	15
Level 1 block totals	*																																		
Level 2 block totals				5	*		0	*	22				48			8*	68				50				46			20	*			59			*
Level 3 block totals				5			0		22					56		118					118				46							79			*
Level 4 block totals				5										78			118								125										*

$A_{....} = 326$ $S = 35$ $A_{....}/S = 9.31$

Divergence at the levels:

$I_1 = 2 \ln 2/9.31 + 3 \ln 3/9.31 + \ldots + 15 \ln 15/9.31 = 133.341$

$I_2 = 5 \ln (5/5)/9.31 + 22 \ln (22/2)/9.31 + \ldots + 59 \ln (59/5)/9.31 = 82.719$

$I_3 = 5 \ln (5/5)/9.31 + 22 \ln (22/2)/9.31 + \ldots + 79 \ln (79/7)/9.31 = 77.020$

$I_4 = 5 \ln (5/8)/9.31 + 78 \ln (78/8)/9.31 + 118 \ln (118/7)/9.31 + 125 \ln (125/12)/9.31 = 74.039$

Level (r)	Divergence (I_r)	Information specific to level (A_r)	Number of blocks (B_r)	Degrees of freedom
1	133.341	50.617	35	25
2	82.719	5.513	10	3
3	77.020	3.175	7	3
4 and higher	74.039	74.039	4	3

A case of ρ_m is presented, using interaction information. $p, p-1, p-2, \ldots, 2$ dimensional interactions are considered. For this A is defined as a p-dimensional dispersion array. The quantity

$$I(1; \ldots; p) \tag{42}$$

measures the interaction among the p variables, defined by extension of (36). After removing variable m, the interaction in the set of $p-1$ variables becomes

$$I(1; \ldots; p-1|m) \tag{43}$$

and the redundancy in the set, conditional on m,

$$\rho_m = I(1; \ldots; p) - I(1; \ldots; p-1|m) . \tag{44}$$

This can in fact be used as a weight for variable m, since it is the information which m shares with the other variables. Since this is the same as the information that the other variables have with m, ρ_m is a symmetric measure of redundancy.

Equivocation

Whereas redundancy is a concept related to interaction, equivocation is a complementary concept related to specificity. Equivocation can be formulated starting from either (3) or (5), leading to weighting systems for individuals, variables, or both.

The measure based on (3) is an equivocation information, given as (21). This is decomposed into components based on (24) and (25) which with (20), the interaction term, form an additive sequence:

Source	Information	Formula
Entity u	$I(A_{u|z})$	(24)
Entity z	$I(A_{z|u})$	(25)
Interaction	$I(A_u;A_z)$	(20)
Multiple of joint entropy	$I(A_u,A_z)$	(19)

$I(A_{u|z})$ and $I(A_{z|u})$ are indeed multiples of conditional entropy. They measure the specific contributions of u and z to the joint entropy. In relative terms:

$$R(A_{u|z}) = 100 \frac{I(A_{u|z})}{I(A_u,A_z)} \%$$

$$R(A_{z|u}) = 100 \frac{I(A_{z|u})}{I(A_u,A_z)} \% . \tag{45}$$

These formulae can be extended to any number of entities.

An equivocation quantity can be derived also based on (5) yielding the components of (37):

Source	Divergence	Formula
Entity u	$I(A_{u|z})$	(34)
Entity z	$I(A_{z|u})$	(35)
Interaction	$I(A_{uz};A_{uz}^0)$	(36)
Joint divergence	$I(A_{u*,z*};\overline{A}_{u.,z.})$	(37)

$I(A_{u|z})$ measures the information contributed specifically by u to the joint divergence. $I(A_{z|u})$ measures the same for z. Similar formulae can be derived to accommodate any numbers of entities.

5. Application to community analysis

Ecological connections

The vegetation system may be viewed from the perspective of landscape, niche, or response. The analytical methods formulated for the specific cases, however, need not to be isolated, since landscape, niche and response are interrelated parts of a complex system. When viewed from a landscape perspective, the spatial pattern of the system comes into focus. The spatial dispersion of a species is a case in point. The niche perspective requires placing the system within the confines of an ecological space. The response perspective leads to focusing on trajectories, and thus, on the niche shape.

One of the earliest and frequently quoted monographs which treats pattern in the vegetation system in a landscape context is Greig-Smith's (1952, 1964). It seeks to define spatial pattern as a departure from random arrangement conditional on quadrat sampling and a sum of squares criterion to measure variation. Greig-Smith's analysis in fact incorporates an analogue of our nested model. The biological theory of pattern and relevant methodology is most completely treated in their full richness by Pielou in numerous papers and several books (1969, 1974, 1975). Specific contributions and overview are discussed by the numerous contributors to the New Haven Symposium collated by Patil, Pielou & Walters (1971). The methods can almost invariably be restated in terms of entropy or divergence measures. One example is given by Orlóci (1971) which is an information theoretical adaptation of Greig-Smith's pattern analysis. Other adaptations are particularly well facilitated for

24

methods using transition probabilities (Pielou 1975 and references) or contingency tables in specific contexts (Juhász-Nagy 1980, Feoli, Feoli-Chiapella, Ganis & Sorge 1980).

Species diversity has been considered an important structural property of multi-species communities. The diversity measures however greatly differ and many may not directly be derived from entropy (Williams 1964, Hurlbert 1971). The diffusion of entropy-based measures in ecology is mainly owing to the work of Margalef (1958) and McArthur (1965). Comprehensive studies on diversity measures have been published (Hill 1973, Pielou 1975, Grassle, Patil, Smith & Tallie 1979) and numerous attempts have been made to find community properties that can be efficiently described by these measures. Community stability (Kushlan 1976, Hurd et al. 1971) niche width and overlap (Petraitis 1973, Hanski 1978), community dynamics (Williams, Lance, Webb, Tracey & Dale 1969) or community homogeneity (Lausi 1972) are cases in point. However, the entropy based measures, most often Shannon's function (3), proved to be not quite up to par with the task (cf. Gallucci 1973) owing to complexities in the system which a single scalar, such as the diversity measure, cannot possibly describe.

Depending on how the individual niches overlap, communities of species are formed and trends of compositional variation evolve. In this process the environmental influence is decisive, since species and communities respond to it. To reveal trends or discontinuities in compositional variation, cluster analysis is one of the methods (Sneath & Sokal 1973, Jardine & Sibson 1971; Anderberg 1973, Clifford & Stephenson 1975; Orlóci 1978a). Cluster analysis structures the sample into groups, which can serve for prediction and identification. In these, and also in many analyses that are routinely performed on classifications, information functions have considerable potential (Orlóci 1978a and references therein). It is the primary goal of the next sections to show this potential and to outline a comprehensive and flexible methodology.

Classification

The process of classification requires class recognition (cluster analysis) and assignment of new individuals to parental classes (identification). In either case similarities or affinities have to be measured based on selected variables. The variables may or may not be weighted.

A. Weighting of variables and individuals

Weights (W) may be given to variables to modify their influence on the analysis (Orlóci 1978a) or to individuals for various reasons (Feoli & Feoli-Chiapella 1979). The weighting procedures differ (cf. Orlóci 1973, 1978a, van der Maarel 1979) and some use information (Orlóci 1976, 1978b, Feoli 1976). The weights are components of interaction divergence or entropy difference.

A weighting algorithm may be formulated in different terms. Redundancy and equivocation are cases in point:

(1) Weighting by equivocation entropy

The weights of the mth of the entities, conditional on the remaining $t-1$,

$$W_m = I(A_m) - \rho(m) \tag{46}$$

where $\rho(m)$ accords with (44) and $I(A_m)$ with (17) or (18) depending whether rows or columns are involved. This is in fact the equivocation entropy for the mth principal marginal distribution of an n-dimensional dispersion array, measuring the uniques of m. The complement of this is ρ_m which as a weight would measure the dependence of m on the other entities in the set.

(2) Weighting by equivocation divergence

In this case, W_m is given by the main effect (34) in the partitioning of the joint effect in an n-dimensional dispersion array, defined by expression (37). This measures how much of the joint effect is specific to entity m (Orlóci 1978b).

(3) Weighting by interaction

Whereas the preceding weighting techniques use the specific contribution of an entity to equivocation or joint effect, other techniques use the contribution to interaction. The weighting is thus by redundancy in the set conditional on the entity. This redundancy is equal to the reduction in the total interaction (44) in the set when m is removed (Orlóci 1976).

(4) Convolution of weights

Having determined weights (W) for variables (i) and individuals (j) in the same system, their convolute weight is $W_i W_j$. The $W_i W_j$ emphasizes different properties depending on the weighting function. If it is a mutual information, and if it is for a diversity array, the indistinctness of the element X_{ij} is emphasized. If it is equivocation, the uniqueness of X_{ij} is emphasized.

B. Data reduction

Considering the high demand on core space in most complex compulations involved in cluster analysis, the problem of data reduction has to be faced. This may require ranking to identify entities that may be removed with the least of consequences on information loss. The weight functions can perform the ranking. Should the last q of t entities with the least weight be removed, the total loss of information would be proportional to

$$\sum_{m=1}^{t-q} W_m \Big/ \sum_{m=1}^{t} W_m$$

where W_m is m's weight.

C. Measurement of resemblance

Only selected cases are considered. In the first case, X_u and X_z identify two columns of X,

X_u	X_z	Total
X_{1u}	X_{1z}	$X_{1.}$
X_{2u}	X_{2z}	$X_{2.}$
.	.	.
X_{pu}	X_{pz}	$X_{p.}$

One resemblance measure expresses the direct resemblance of X_u and X_z as a similarity,

$$S_{uz} = - \sum_{j=1}^{p} \sum_{l=u}^{z} \left[X_{jl} \ln \frac{X_{jl}}{X_{j.}} \right]. \tag{47}$$

This is entropy within rows summed for p rows (24). Expansion to more than two columns (or two groups) is self-evident. Independence of rows is assumed. S_{uz} is maximal when $X_{ju} = X_{jz}$ for all j. To relativize S_{uz}, it can be expressed as a ratio of the maximum:

$$\max S_{uz} = \sum_{j=1}^{p} X_{j.} \ln 2 . \tag{48}$$

A divergence measure on X_u, X_z is

$$D_{uz} = \sum_{j=1}^{p} \sum_{l=u}^{z} X_{il} \ln \frac{2X_{il}}{X_{j.}} = \max S_{uz} - S_{uz} \tag{49}$$

This is a dissimilarity, analogous to (34). To relativize D_{uz}, it can be divided by the maximum which is the same as (48). This occurs when X_u and X_z have nothing in common. At that state, all entropy is in divergence (negentropy) and there is no entropy in evenness.

The elements in X may represent arbitrary symbols, such as ., +, l, etc. If in row j of p rows two columns agree, their divergence (38) is zero. If different, the divergence is maximal, $2 \ln 2$. For the p rows, the maximum is $2p \ln 2$. For this reason, for two columns X_u, X_z,

$$D_{uz} = \sum_{j=1}^{p} \left[D_{uz|j} = \begin{cases} 0 \text{ if } X_{ju} \neq X_{jz} \\ \ln 4 \text{ if } X_{ju} = X_{jz} \end{cases} \right]. \tag{50}$$

When (50) is divided by $2p \ln 2$, the divergence is expressed as a relative quantity. In this case D_{uz} measures similarity, since it operates on p two-valued frequency distributions into which the X_{ju}, X_{jz} are condensed.

The pair X_u and X_z may be treated as a $p \times 2$ contingency table in the case of a diversity array. A dissimilarity measure then is the quantity,

$$I(X_u; X_z) = \sum_{j=1}^{p} \sum_{l=u}^{z} X_{jl} \ln \frac{X_{jl}(X_{.u} + X_{.z})}{X_{j.}X_{.l}}. \tag{51}$$

This is analogous to (36) and expandable to groups in the form of (30, 39).

Another formula for measuring resemblance when binary data are used is suggested by Williams, Lambert & Lance (1966, Lance & Williams 1968). It is an entropy measure for diversity arrays,

$$I = p n \ln n - \sum_{j=1}^{p} [a_j \ln a_j + (n - a_j) \ln (n - a_j)] \tag{52}$$

where n is the number of objects, p is the number of variables, a_j the number of objects having variable j and $n - a_j$ is the number of objects lacking it. (52) is a zero quantity when the set is completely homogeneous. The analogy with (17) and (18) is obvious. Other formulae based on entropy as a diversity measure have been discussed by Rejmánek (1981).

Lagonegro and Feoli (1979) proposed a similarity measure which they defined as relativized mutual information in diversity arrays,

$$S_{uz} = 1 - I(A_{uz}; A_{uz}^0)/(A_{..} \ln p). \tag{53}$$

In a dispersion array, such as a co-occurrence matrix for species, (53) measure relative redundancy. If $S_{uz} = 1$, species do not carry specific information. Numerical examples are given in Table 7.

D. Cluster-seeking algorithms

Clustering algorithms employ two main strategies: monothetic or polythetic. Both can be agglomerative and divisive (Williams 1971, Cormack 1971). Clustering algorithms use mainly resemblance matrices. These can be generated by the functions already discussed. A monothetic method has been suggested by Lance and Williams (1968). The decision to subdivide any set on species j is conditional on maximizing the difference,

$$\Delta I = I(J^+ + J^-) - I(J^+) - I(J^-). \tag{54}$$

In this, J^+ is the set of relevés with species j, and J^- is the set without j. $(J^+ + J^-)$ is the union set. The individual terms of I accord with (52). Bottomley (1971) com-

Table 7. Examples involving the application of formulae 47, 48, 49, 51, 52 and 53 based on relevés of Table 2.

Formula (47)

$$S_{uz} = -(2 \ln 1/2 + 5 \ln 5/9 + \ldots + 10 \ln 10/30) = 117.58$$

Formula (48)

$$\max S_{uz} = 4 \ln 2 + 7 \ln 2 + \ldots + 30 \ln 2 + 13 \ln 2 = 212.6$$

Formula (49)

$$D_{uz} = 4 \ln 2 + 7 \ln 2 + \ldots + 13 \ln 2 = 95.02$$

Formula (51)

$$X_{.1} = 189$$

$$X_{.2} = 118$$

$$X_{..} = 307$$

$$I(X_1;X_2) = 4 \ln((4)(307))/((118)(4)) + 7 \ln((7)(307))/(189)(7)) + \ldots +$$
$$3 \ln((13)(307))/((118)(13)) = 76.022$$

Formula (52)

$$I = 34 (2) \ln 2 - 14 (2) \ln 2 = 47.134 - 19.408 = 27.726$$

Formula (53)

$$S_{uz} = 1 - 76.022/307 \ln 34 = 1 - 0.070 = 0.93$$

mented on the procedure and algorithms are available from the authors.

Another algorithm has been proposed by Podani (1973) as a version of association analysis (Williams & Lambert 1959). The decision rule is specified by

$$I = \max \sum_{u \neq z}^{p} I(A_u;A_z) \tag{55}$$

where $I(A_u;A_z)$ is the interaction information in an R-dispersion array in which species vectors are considered in pairs (20, 36, 51).

Polythetic agglomerative methods are numerous, and sufficient reviews are found in Cormack (1971), Williams (1971), Anderberg (1973), Sneath & Sokal (1973), Bock (1973), Hartigan (1975) and Orlóci (1978). The algorithms are either data based or coefficient based. The data based algorithms (stored data approach) require continued access to the data. The coefficient based algorithms require no access to the data.

In the data based methods, the clustering is usually based on

$$d(A;B) = I_{(A+B)} - I_A - I_B \tag{56}$$

(Orlóci, 1969a,b, 1970a,b, 1972a,b) which is either minimized (agglomerative clustering) or maximized (subdivisive clustering). A and B are disjoint subsets each containing at least one individual. Implicit in (56) is the inequality

$$I_{(A+B)} \geqslant I_A + I_B .\tag{57}$$

$I_{(A+B)}$, $I_{(A)}$ and $I_{(B)}$ can be computed by using different formulae of entropy or divergence which were discussed in Section 4. If criterion (56) is chosen, after each fusion or subdivision the quantities $I_{(A+B)}$, $I_{(A)}$ and $I_{(B)}$ must be recomputed from the data. It is to be kept in mind, however, that the data based algorithms can supply an additive measure of classification efficiency. For example, at the k-group level in a hierarchy,

Source	Information	Efficiency
Between groups	$I_B = I_T - I_W$	I_B/I_T
Within groups	$I_W = I_1 + I_2 + \ldots + I_k$	
Total	I_T	

If a coefficient based algorithm is used, classification efficiency has to be defined differently.

Predictivity analysis

The predictive value of a classification and its efficiency are related. Predictivity is measurable based on internal or external criteria. When internal criteria are used, predictive value means the ability of a classification to predict the presence of a species, a specific state, or combinations within the classes into which a multitude is subdivided. Based on external criteria predictivity is a function of correlations with the states or sets of states of external variables.

A measure of predictivity is the interaction information as formulated for dispersion arrays (20). For te ith class, corresponding to a class in a given classification U, the interaction information shared with classification Z is

$$I(U_i;Z) = \sum_{j=1}^{s_z} A_{ui,zj} \ln \frac{A_{ui,zj} A_{u.,z.}}{A_{ui,z.} A_{u.,zj}} .\tag{58}$$

U_i may be a class in a vegetation classification and Z may be an environmental classification. This would facilitate external prediction. U_i may be defined as a species group and Z as a vegetation classification. In this case, the prediction would be internal.

To find the level in a hierarchy at which a classification is most predictive, interaction information is best expressed in relative terms. Formulae (22) and (23) are suitable for this, but other formulae could be considered, e.g., (53). Another may be

$$D(A_u;A_z) = (1 - d(A_u;A_z))/(1 + d(A_u;A_z)) \tag{59}$$

(Feoli & Lausi 1980). The $d(A_u;A_z)$ is defined as (22). Formula (59) is more sensitive than (22) at extreme divergence. Other measures are readily conceived in connection with, for instance, the nested models.

Comparison of classifications

Classifications of the same data set obtained by two different methods can be compared. The interaction information (36) is a relevant criterion. The elements in the dispersion matrix indicate joint possessions of units between the classes of the two classifications. The higher the interaction information, the more similar are the classifications. This indicates high predictivity.

Identification

The process of finding a parental class for a new object in a given classification is called identification. An object is assigned to a class to which it has the highest affinity. The fusion criteria of the clustering methods may be a basis for identification. Upon completion of identification, the question of reliability should be addressed. For this, procedures similar to classification efficiency may be used, or some Bayesian analysis may be applied. Reference is made to formulations in the preceding sections and Orlóci (1978a).

6. Computer programs and examples of application

The methods are programmed and illustrated by worked examples. A data set (Table B1) of 35 quadrats of grasslands from the Friuli Region of Italy is used. The data elements are occupancy counts for species in quadrats of 1 metre square size, subdivided into 25 subquadrats. The number of species in the data set is 108. The description of programs are given as follows:

A. Characterizations of programs

Seven programs of information analysis are presented. The programs perform the analyses discussed in the main text. The following is a brief review:

Program	Analysis performed

FRANKI — Weights are determined for individuals (relevés) or variables (species). The criterion is interaction information or equivocation divergence. The data set is an array of alpha numeric symbols. Dispersion matrices are computed within the program. Option for printing the data table is included. The weights are ordered and printed with residuals.

SINFUN* — The different subroutines compute either resemblance values or other quantities which are useful in the evaluation of table homogeneity/heterogeneity, comparison of classifications, derivation of weights for variables or individuals, prediction, and identification. The subroutines use data packing and handle large data sets. For technical details regarding SINFUN see Lagonegro & Feoli (1981) and Table A1. When a file of the results is written, the file name is LOUTDIS and the format is (5X,10G12.5), involving the lower triangle of the resemblance matrix.

NESTOFL — Two options are offered as described in section F. Option NESTYP1 computes a nested analysis, according to quantities defined by the formulae in Table 4. Option NESTYPE2 computes interaction information (36) for a given species and classes, at any specified level in a hierarchical classification of relevés. Data input is from a diversity array.

IAHOPA* — Data sets are structured, subsets are isolated and compared or tested for homogeneity as requested. Four branches are included. These are described by Lagonegro & Feoli (1979), Feoli & Lagonegro (1982) and in Table A2. The input data set is a diversity array.

The following *clustering programs* are all based on minimization of criterion (56), but with different information functions involved and with different types of arrays. The algorithms are data based and produce hierarchical classifications:

CINF1F — Data input is from an alphanumeric array, the rows of which are condensed in the program into marginal dispersion arrays on which (38) is computed. If continuous variables are input, the data are categorized according to interval classes defined internally within the program.

* Programs SINFUN and IAHOPA use system library subroutine MOVCHR (N,A,M,B). This moves the Nth byte of vector or word A into the Mth byte of vector or word B. Subroutine MOVCHR is used for packing (unpacking) data.

32

Table A1. Options of SINFUN. The input data are indicated by X. Arrays created by the programs are indicated by C.

Option	Data input			Quantity computed/ entities involved	Function/ criterion	Remarks
	Alpha-numeric	Diversity array	Dispersion array			
INF1		X		Resemblance/ column groups	(47)	Pairwise comparisons using entropy
INF1F	X		C	Resemblance/ column groups	(28)	Pairwise comparisons using information
INF2		X		Resemblance/ column groups	(36)	Pairwise comparisons using interaction information
INF3		X		Resemblance/ individual columns	(21)	Pairwise comparisons using equivocation information
INF3F	X		C	Resemblance/individual columns	(21)	Pairwise comparisons using equivocation information
DAB		X		Resemblance/ column groups	(34)	Pairwise comparisons using row main effects
SINGLEHET		X or	X	Homogeneity/ single array	(37)	Homogeneity tested within individual rows
CLACOMP		X or	X	Decomposition of joint information (19)/a single array	(17), (18), (20) and (17), (24) and (18), (25) and (58)	Results useful in comparison of classifications, weighting, and predictive analysis
DISCAN		X		Identification/individual column	(34)	Individuals are assigned to groups, using a nearest neighbour criterion

Table A2. Branches of IAHOPA.

Branch	Operations	Criterion
MLTAX1*	Data set rearranged according to sequence of species based on a nearest neighbour criterion applied to a species association matrix	Association matrix computed according to (22) applied to diversity arrays. Arbitrary sequence of species can be specified
SCRAPS	Subsets isolated in data tables and organized for comparisons	Bounds for subsets specified by user
MLTAX2	Homogeneity tested within subsets isolated in SCRAPS or given from an external input file	(36) and (53)
MLTAX3	Subsets isolated in SCRAPS or given data tables compared in pairs	(36) and (53)

* MLTAX1 is an expanded version of an earlier program by Feoli & Lagonegro (1979) and Lagonegro & Feoli (1979) under same name.

CINF2	The data input is from a diversity array on which function (36) is applied directly.
CDAB	A diversity array is the input. The information function is (34).

B. Ranking species

The species of Table B1 are used as example. The weights are interaction information quantities (FRANKI, option ID7 = 0). The list is truncated by deleting species in the lower 40%. The total interaction information for 108 species is 3309.7. To find cumulative percentages divide the total minus residual by total.

To run FRANKI, the following cards are required:

Card 1 – Title of run and specifications for NP, NQ, ID7, INF, ITR, IPR

```
            40      45     50     55     60     65     70     75     80
            I       I      I      I      I      I      I      I      I
RANKING SPECIES   108     35      0      5      0      1
```

Card 2 – Input format for data (INPFMT)

```
      10          20
      I           I
   (35I2)
```

Card 3 and following cards – data in format (35I2)

```
 2    4    6    8   10   12   14   16   18  .................................................
--I---I---I---I---I---I---I---I---I---I---I---I---I---I---I ---------
 0    0    0    0    0    0   15    0    0  .................................................
..................................................................................................
..................................................................................................
..................................................................................................
-----------------------------------------------------------------------------------------
```

Card 4 – Left blank

```
-----------------------------------------------------------------------------------------
-----------------------------------------------------------------------------------------
```

Abbreviations:

NP = number of species
NQ = number of relevés
ID7 = 0, 1, or 2 (option for ranking based on interaction, equivocation divergence, or Rajski's metric)
ITR = 0 or >0 (option for transposition; 0 for species weights, >0 for relevé weights)
IPR = 0 or 1 (0 for no printing of data table, 1 for printing)
INF = 5 or 1 (for input from cards, or tape 1).

C. Computation of resemblance matrices

Resemblance matrices are computed by program SINFUN. The following cards are needed:

Card 1 – Title of run and specifications NTB, INF, IFUNCT, NDPW, NROWS, LOUTDIS

```
               30      35    40              50    55   60    65          
------------------I-----I-----I-------------I-----I----I-----I------I-----I----
INFIF-SI          -35          5INF1F              1    34
```

Card 2 – Input format for data and specifications NDIG, NDEC

```
                                            70     75    80
(35F2.0)                                            2     0
```

Card 3 – Format specification for printing data table and IND

```
                                            70     75    80
(3X,35F3.0)
```

Table B1. Data table. Contents extracted from published table by Feoli, Parente & Trinco (1979).

Species	Quadrat 1	2	3	4	5	6	7	8	9	10	11	12	13	14	15	16	17	18	19	20	21	22	23	24	25	26	27	28	29	30	31	32	33	34	35
1. Helictotrichon pratense (L.) Pilger	25	10	5	10	3	·	·	·	·	·	·	·	·	·	·	·	2	·	·	·	·	·	·	·	·	·	·	·	·	·	·	·	3	·	·
2. Agrostis tenuis Sibth.	20	6	6	25	12	8	·	·	9	15	4	6	2	·	·	·	·	·	7	·	·	22	9	4	6	·	18	·	20	1	12	12	8	10	·
3. Anthoxanthum odoratum L.	3	2	·	4	·	·	·	·	·	·	·	7	7	2	·	·	·	·	·	·	·	20	8	12	·	25	22	·	·	2	6	·	·	·	·
4. Arrhenatherum elatius (L.) JK Presl	5	5	·	10	2	5	4	2	20	3	7	2	·	3	2	·	4	·	3	·	3	3	7	8	10	3	10	3	5	12	6	10	3	3	4
5. Brachypodium pinnatum (L.) Beauv.	·	·	·	·	·	·	3	23	·	3	8	2	6	2	4	·	4	4	·	·	·	4	·	12	·	·	·	·	9	5	3	5	4	4	6
6. Briza media L.	4	5	·	·	·	·	·	·	·	8	18	·	·	7	4	·	4	4	10	·	·	·	·	10	·	·	·	·	10	4	3	5	4	8	4
7. Bromus erectus Huds	20	10	5	·	·	·	17	3	·	8	18	8	8	18	3	·	4	5	4	·	2	6	5	20	10	15	15	9	10	4	6	25	25	15	24
8. Chrysopogon gryllus (L.) Trin.	·	5	2	·	·	·	4	·	·	8	·	·	·	·	·	·	·	·	3	·	·	4	3	3	·	·	5	5	·	4	·	·	·	3	6
9. Cynosurus cristatus L.	7	3	·	·	·	·	·	·	·	·	·	·	·	·	·	·	·	·	·	·	·	3	·	·	·	·	·	2	2	·	·	·	·	3	·
10. Carex flacca Schreb.	10	·	·	·	·	·	·	·	·	·	·	·	2	·	3	8	·	·	·	·	·	·	·	·	8	·	·	·	·	1	·	·	·	·	6
11. Carex montana L.	·	6	12	·	5	·	·	·	·	15	6	12	15	·	18	·	8	6	·	·	20	·	·	·	1	5	·	·	10	8	10	23	12	·	8
12. Carex caryophyllea Latourr.	6	5	·	·	·	·	·	·	·	·	·	1	·	·	18	·	·	·	9	4	·	·	·	·	·	5	·	15	1	·	6	·	·	·	·
13. Dactylis glomerata L.	·	·	·	·	15	10	7	·	·	2	·	2	1	·	16	·	·	3	·	·	·	10	2	8	7	5	3	12	10	2	6	·	1	2	4
14. Danthonia alpina Vest	·	·	5	10	3	·	10	20	·	7	11	6	15	9	·	·	3	11	6	12	·	·	·	·	1	·	·	·	2	·	·	·	12	2	·
15. Danthonia decumbens (L.) DC.	·	4	3	15	15	·	·	15	3	18	16	4	8	16	6	·	8	8	20	16	12	6	·	5	3	4	15	7	16	12	10	5	3	5	4
16. Festuca tenuifolia Sibth.	·	5	2	·	4	·	7	·	18	18	4	·	7	·	18	5	23	10	15	·	5	·	7	5	3	3	21	·	15	·	5	·	5	6	8
17. Festuca rubra L.	·	·	8	·	·	·	·	4	4	3	10	5	·	1	·	·	·	·	5	·	·	20	2	·	·	·	·	·	·	·	13	·	·	·	5
18. Festuca pratensis Huds.	12	7	·	·	·	2	2	·	2	7	4	·	·	3	4	3	4	6	3	3	3	10	10	·	4	5	9	2	10	1	4	12	12	6	4
19. Holcus lanatus L.	·	·	3	·	·	·	·	12	·	·	1	5	·	·	·	·	·	5	·	·	·	·	4	4	·	·	·	15	·	·	3	·	·	·	·
20. Koeleria pyramidata Auct.	·	·	·	10	4	25	·	1	4	·	·	·	·	3	2	2	·	4	·	·	·	·	·	·	·	·	·	·	4	·	·	4	·	·	·
21. Trisetum flavescens (L.) Beauv.	·	·	·	·	3	3	·	·	5	5	8	·	20	6	·	15	·	·	·	10	·	·	·	6	·	·	·	6	·	·	·	·	·	·	·
22. Molinia arundinacea Schrank	·	3	·	·	·	1	·	·	·	·	·	·	·	·	·	·	·	·	·	·	·	·	·	·	3	·	·	·	3	·	·	·	·	·	·
23. Nardus stricta L.	·	·	·	·	·	·	·	·	·	·	·	·	·	·	·	·	·	·	·	·	·	·	·	·	·	·	·	·	·	·	·	·	·	·	·
24. Poa pratensis L. & P. trivialis L.	·	1	·	·	·	·	·	·	·	·	·	·	·	·	·	·	·	5	·	·	·	·	·	·	·	·	·	6	·	3	·	·	·	·	·
25. Carum carvi L.	·	·	1	·	·	·	·	·	·	·	·	·	·	·	·	·	·	4	3	·	·	·	·	·	·	·	·	·	·	·	·	·	·	·	·
26. Carex pallescens L.	·	·	·	·	8	1	·	·	·	·	·	·	·	·	·	·	·	·	·	·	·	·	2	·	·	·	·	·	9	·	2	·	·	·	·
27. Luzula multiflora (Ehrh.) Lej.	3	·	·	·	·	6	·	1	4	1	·	·	·	·	2	·	·	·	·	·	·	·	10	·	·	·	·	·	·	·	·	2	3	·	·
28. Anthyllis vulneraria L.	·	·	·	·	·	·	·	·	·	·	·	·	·	·	·	·	·	·	·	·	·	·	·	·	6	·	·	·	·	·	3	·	2	3	1
29. Genista tinctoria L.	·	·	·	·	4	2	·	2	·	5	2	·	3	·	·	·	2	·	2	2	·	5	5	·	·	·	22	·	·	·	·	1	·	·	18
30. Lathyrus pratensis L.	7	·	1	·	·	·	·	·	3	2	·	·	5	·	·	·	·	2	2	·	·	7	7	5	·	·	·	·	·	·	1	1	·	·	·
31. Lotus corniculatus L.	4	6	·	4	·	4	·	·	·	6	6	7	·	6	10	·	·	·	3	1	4	4	·	2	8	7	6	10	10	6	·	5	5	8	4
32. Ononis spinosa L.	·	1	·	·	·	·	·	·	·	·	·	·	·	2	2	·	·	·	·	·	·	·	·	·	·	·	·	3	·	·	·	·	·	·	·
33. Onobrychis arenaria DC.	·	·	·	·	·	·	·	·	·	·	·	·	·	·	·	·	·	·	·	·	·	·	·	·	·	13	·	·	·	3	·	·	·	·	·
34. Trifolium campestre Schreb.	1	·	3	·	·	·	·	·	·	·	·	·	·	·	·	·	·	·	·	·	·	·	·	·	·	1	1	2	9	·	2	1	·	·	·
35. Trifolium montanum L.	1	1	·	·	·	2	2	·	5	4	5	·	2	3	·	·	·	·	3	3	·	3	2	·	5	5	·	·	10	1	2	3	8	·	·
36. Trifolium pratense L.	7	3	3	22	·	·	·	·	1	5	·	·	·	·	·	6	·	2	5	·	·	8	4	·	·	·	6	10	·	6	3	4	4	5	·

Table B1 (continued).

Species	1	2	3	4	5	6	7	8	9	10	11	12	13	14	15	16	17	18	19	20	21	22	23	24	25	26	27	28	29	30	31	32	33	34	35
37. Trifolium repens L.	6	.	.	.	.	.	.	.	.	.	.	.	.	.	.	.	.	.	.	.	.	.	.	.	.	3	3	.	.	.	.	.	.	.	.
38. Trifolium rubens L.	.	.	.	.	.	.	.	.	.	.	.	.	.	.	.	.	.	.	.	.	.	.	.	.	.	.	.	.	.	.	2	.	.	1	.
39. Ajuga reptans L.	5	.	.	.	.	.	.	.	.	.	.	1	.	.	.	.	.	.	.	.	2	.	.	.	.	.	.	.	.	7	2	3	1	.	.
40. Prunella laciniata (L.) L.	.	1	1	.	.	.	.	.	.	.	.	4	.	.	.	5	.	.	.	.	.	.	.	.	.	.	.	.	.	.	.	.	3	6	.
41. Salvia pratensis L.	.	1	.	.	4	.	.	.	.	.	1	.	4	.	.	.	.	.	.	.	.	.	.	.	.	1	.	4	.	1	.	1	.	.	.
42. Arnica montana L.	.	.	.	.	4	.	1	2	.	6	.	.	.	.	.	1	.	9	4	15	.	.	.	.	.	4	.	4	.	.	.	.	.	.	.
43. Hypochoeris maculata L.	.	.	1	.	.	6	.	.	.	6	.	.	.	.	.	1	.	2	.	.	.	.	.	.	.	.	.	.	.	.	.	.	.	.	.
44. Buphthalmum salicifolium L.	.	5	.	.	2	.	.	.	.	.	.	6	.	.	.	4	.	4	.	.	.	.	.	.	5	2	.	.	.	.	.	.	.	.	.
45. Betonica officinalis L.	.	.	2	6	2	9	.	2	.	6	.	.	3	.	.	4	.	.	1	.	3	.	.	.	2	.	.	.	.	.	2	.	2	.	.
46. Campanula glomerata L.	10	.	3	.	2	2	.	5	.	.	.	.	.	.	.	.	.	.	2	.	.	.	.	.	.	.	.	.	3	.	.	.	1	3	.
47. Polygala nicaensis Risso	.	.	3	.	.	2	.	.	1	.	.	3	.	.	.	.	.	.	.	.	.	.	.	.	.	.	.	.	.	.	.	.	.	.	.
48. Campanula scheuckzeri Vill.	.	.	.	.	2	.	.	3	.	.	2	.	.	.	.	2	.	.	.	.	.	.	.	.	.	.	.	.	.	.	1	.	.	.	.
49. Leucanthemum vulgare Lam.	.	.	4	4	2	.	.	.	.	8	.	.	.	.	9	2	.	3	.	.	.	.	.	.	.	.	.	.	6	.	.	.	1	.	1
50. Carlina vulgaris L.	.	.	.	.	.	.	.	.	.	.	.	.	1	.	9	2	.	2	.	.	1	.	.	.	.	.	.	3	.	.	1	.	.	.	.
51. Centaurea scabiosa L.	.	.	.	.	.	.	.	.	.	.	.	.	.	.	.	.	.	.	.	.	.	.	.	.	.	.	1	.	1	.	.	.	.	.	.
52. Cirsium pannonicum (L.) L.K.	.	2	2	.	.	1	1	.	.	.	.	.	.	.	8	.	.	.	.	.	8	.	.	8	.	2	.	.	.	2	4	4	.	1	.
53. Cruciata glabra (L.) Ehrend.	20	.	7	.	.	8	8	.	.	6	15	5	9	.	.	.	.	12	.	.	.	.	18	10	15	8	.	.	.	4	4	4	.	2	.
54. Calluna vulgaris (L.) Hull	.	.	.	20	.	.	.	25	25	12	.	6	6	.	8	.	.	.	.	.	.	.	.	.	.	.	.	.	.	.	.	.	.	.	.
55. Serratula tinctoria L.	10	.	.	.	.	.	.	.	.	.	.	.	.	.	.	.	.	.	.	1	.	.	.	.	.	.	.	.	.	.	.	.	.	.	.
56. Chamaecytisus hirsutus (L.) LK.	.	.	2	2	.	5	3	.	.	.	.	.	.	2	.	.	.	.	.	.	.	.	.	.	2	.	.	.	.	.	.	.	.	.	4
57. Polygala comosa Schbsuhr	.	.	.	2	.	1	.	.	.	.	.	.	1	.	.	.	.	.	.	.	2	.	.	3	.	.	.	.	.	.	.	.	.	3	.
58. Adenophora lilifolia (L.) A. DC.	.	.	.	.	.	.	1	.	.	.	.	.	.	.	1	.	.	.	.	.	.	1	.	2	.	.	.	.	.	.	.	.	.	.	.
59. Dianthus carthusianorum L.	.	.	.	.	.	.	.	.	.	.	.	.	.	.	.	.	.	.	.	.	.	.	.	.	.	1	.	.	.	.	5	.	.	.	.
60. Euphorbia verrucosa L.	.	.	4	.	.	.	2	.	.	3	.	.	.	.	.	.	.	.	7	.	.	.	.	.	.	.	.	.	.	.	.	.	.	.	.
61. Euphorbia cyparissias L.	.	.	.	.	.	.	2	.	.	.	.	.	.	.	8	.	.	.	.	.	.	.	.	8	.	.	.	.	.	4	.	.	2	1	1
62. Cerastium arvense L.	.	.	.	.	.	.	.	.	.	.	.	5	.	.	.	.	.	.	.	.	.	.	.	.	.	.	3	.	.	4	.	12	.	.	.
63. Gratiola officinalis L.	.	.	.	.	.	.	.	.	.	.	.	.	.	5	.	.	.	.	.	1	.	.	.	.	.	.	.	.	.	.	.	.	.	.	.
64. Galium mollugo L.	.	1	1	.	.	.	.	.	.	.	.	.	3	.	.	.	.	.	.	.	.	.	.	.	.	6	.	.	.	.	.	.	.	.	1
65. Galium verum L.	1	7	.	.	1	.	1	.	.	.	.	.	.	.	.	5	.	.	.	.	6	3	8	25	10	.	3	.	.	1	4	.	3	.	2
66. Galium lucidum All.	.	.	.	.	.	.	2	.	4	4	5	.	.	.	.	.	.	6	6	.	6	.	.	.	.	.	3	3	15	.	.	.	.	5	2
67. Gentiana kochiana Perr. et Song.	.	.	.	.	.	.	.	.	.	.	.	.	.	.	.	.	.	1	3	.	.	.	8	.	.	.	.	1	.	.	.	.	.	1	.
68. Gladiolus palustris Gaudin	.	.	.	.	.	.	.	.	.	.	.	.	.	.	.	.	.	.	.	.	.	.	.	.	.	.	.	.	.	.	2	.	.	.	.
69. Anthericum ramosum L.	.	.	.	.	.	1	.	.	.	.	.	.	.	.	.	.	.	.	.	.	.	.	.	.	.	.	.	.	.	.	.	.	.	.	.
70. Colchicum autumnale L.	2	.	.	.	.	.	.	.	2	.	.	.	.	.	.	.	.	.	.	.	.	.	2	.	.	.	.	.	.	.	.	.	.	.	.
71. Allium carinatum L.	.	.	2	.	.	.	2	.	.	.	.	.	.	.	.	.	.	.	.	.	.	.	.	.	.	.	.	.	.	.	.	.	3	.	.
72. Helianthemum ovatum (Viv.) Dunal	.	.	.	.	.	.	5	.	.	.	.	4	5	.	.	.	.	.	.	.	.	.	2	.	.	.	.	.	.	.	4	.	3	.	.
73. Hieracium pilosella L.	.	.	.	.	.	.	.	.	6	.	5	.	.	.	.	.	.	.	8	2	.	.	.	.	.	.	.	.	.	.	4	.	.	.	.

Table B1 (continued).

Species	1	2	3	4	5	6	7	8	9	10	11	12	13	14	15	16	17	18	19	20	21	22	23	24	25	26	27	28	29	30	31	32	33	34	35
																																			Quadrat
74. Hypericum perforatum L.	.	.	.	.	.	.	.	3	.	.	.	.	.	.	.	.	.	.	.	.	.	.	.	.	.	.	.	.	.	.	2	2	.	.	2
75. Linum catharticum L.	.	2	3	.	.	.	6	.	.	.	.	.	.	.	.	6	.	.	.	.	.	.	.	.	.	2	3	.	.	1	2	10	.	.	.
76. Knautia drymeia Heuff.	5	4	4	10	.	.	6	.	2	.	3	3	.	.	2	7	.	4	7	.	.	5	.	4	7	3	.	.	2	.	.	10	3	15	3
77. Leontodon hispidus L.	20	5	12	10	4	.	4	4	2	8	4	12	5	2	5	5	.	.	16	16	16	5	2	18	18	18	11	.	8	.	2	.	15	15	2
78. Centaurea gaudinii (Boiss. Reut.) Gremli	4	5	.	.	.	.	.	.	.	5	.	2	.	.	.	.	.	2	1	.	.	.	.	.	.	.	.	.	1	5	5	.	4	.	.
79. Tragopogon orientalis L.	1	3	.	.	.	.	.	.	.	.	.	.	.	.	.	1	.	.	1	1	.	2	.	.	.	.	.	.	1	1	.	.	.	1	.
80. Orchis morio L.	.	.	.	.	.	.	.	.	.	.	.	.	.	.	.	.	.	.	.	.	.	.	.	.	.	.	.	.	2	.	.	.	.	1	.
81. Peucedanum oreoselinum (L.) Moench	22	3	.	8	8	2	6	6	3	15	7	2	4	.	8	8	2	.	8	6	20	20	1	.	.	2	2	1	10	3	13	.	.	.	.
82. Plantago media L.	.	.	2	2	.	.	.	.	.	.	.	3	.	.	5	5	.	.	.	.	.	.	.	.	.	2	2	3	.	.	.	.	.	.	.
83. Plantago lanceolata L.	4	3	8	3	.	.	7	.	.	8	9	3	.	6	5	5	3	3	16	.	3	12	3	4	8	3	5	1	7	4	12	4	.	8	.
84. Pimpinella major (L.) Huds.	.	.	.	.	.	.	.	.	4	.	3	.	.	.	.	6	6	.	.	.	.	.	.	.	.	15	7	.	.	.	.	.	.	.	2
85. Achillea millefolium L.	2	.	2	2	.	.	.	.	2	.	.	.	.	.	.	.	.	.	2	2	2	2	.	.	.	.	.	.	.	.	1	.	.	.	.
86. Linum viscosum L.	.	.	.	.	.	.	.	.	.	.	.	.	.	.	.	.	.	.	2	.	.	.	.	.	.	.	.	.	.	.	3	3	.	.	.
87. Potentilla alba L.	.	.	.	.	.	.	14	.	.	16	.	.	.	.	.	.	.	.	.	.	10	10	10	.	.	.	.	.	.	2	2	2	.	.	.
88. Potentilla erecta (L.) Rauschel	10	10	.	18	18	6	5	12	18	18	8	12	.	20	.	7	3	16	13	6	15	15	.	.	.	.	.	.	.	.	8	.	6	.	.
89. Gypsophila repens L.	.	.	.	.	.	.	.	.	.	10	.	.	.	.	.	.	1	.	.	.	.	.	.	.	.	.	.	1	.	.	.	.	.	.	.
90. Prunella vulgaris L.	2	3	.	.	.	.	2	.	6	3	.	.	2	.	.	.	4	.	2	.	.	4	.	7	.	.	5	.	.	3	3	.	.	.	.
91. Primula vulgaris Huds.	17	1	.	.	.	.	.	.	.	.	.	2	.	.	.	.	.	3	3	.	.	.	.	.	.	.	5	.	.	.	1	.	.	.	.
92. Phyteuma zahlruckneri Vest	.	.	.	.	.	.	.	.	.	.	.	.	.	.	.	.	.	.	5	.	.	.	.	.	.	.	.	.	.	.	2	.	.	.	.
93. Ranunculus bulbosus L.	.	.	15	.	.	.	.	.	.	.	.	.	.	.	.	8	.	.	.	.	.	.	3	.	.	.	8	8	4	.	.	3	.	.	.
94. Ranunculus memorosus DC.	.	.	.	.	3	.	.	4	4	4	.	.	6	.	.	.	.	.	.	.	.	.	3	4	.	.	.	.	4	.	.	.	.	.	.
95. Ranunculus acris L.	12	1	.	.	.	.	.	12	.	.	.	.	.	.	.	.	.	.	.	.	.	.	.	.	.	.	.	.	3	.	.	.	.	5	5
96. Rhynanthus freynii (Sterneck) Fiori	.	13	.	20	.	.	6	.	5	17	5	18	8	.	.	.	.	5	1	.	25	.	.	12	1	8	.	.	.	.	22	12	23	2	.
97. Rhinanthus glacialis Personn.	.	.	.	.	.	.	.	.	.	.	.	.	.	.	23	.	.	.	.	.	.	.	.	.	.	.	.	.	.	.	.	.	.	.	9
98. Rumex acetosa L.	.	.	.	1	.	.	.	.	.	.	.	.	3	.	.	.	.	2	.	.	.	4	2	.	.	.	4	.	.	1	.	.	.	.	.
99. Vicia cracca L.	.	2	.	.	.	.	.	.	.	.	.	.	.	.	.	.	.	.	1	.	.	.	.	3	.	.	.	1	1	.	.	.	1	.	.
100. Scabiosa columbaria L.	.	.	.	.	.	.	.	.	.	.	.	.	.	.	.	.	.	.	.	.	.	3	.	.	2	.	.	.	.	5	5	.	.	3	4
101. Scorzonera villosa Scop.	3	.	.	.	.	.	.	.	.	.	.	.	.	.	4	.	.	.	.	.	.	2	.	.	.	.	.	.	.	.	6	.	.	.	.
102. Sedum acre L.	.	.	.	.	.	.	.	.	.	.	.	.	.	.	.	.	.	.	.	.	.	.	.	.	.	2	.	.	.	.	.	.	.	.	.
103. Globularia cordifolia L.	.	.	.	.	.	.	.	.	.	.	.	.	.	.	.	.	1	.	.	.	.	.	.	.	.	4	.	.	.	.	.	.	.	.	4
104. Thymus longicaulis K. Presl.	6	4	.	.	.	.	.	.	2	2	.	5	.	.	.	.	.	3	2	.	2	.	.	.	.	1	.	.	.	.	.	.	.	.	.
105. Daucus carota L.	1	.	.	.	.	.	.	.	1	.	.	.	.	.	.	.	.	.	.	.	.	5	.	3	.	.	.	.	.	.	.	.	.	1	.
106. Veronica chamaedrys L.	3	3	.	.	.	.	.	3	6	.	2	.	.	.	.	.	.	.	3	.	.	.	.	2	.	.	5	.	.	.	.	.	.	.	.
107. Viola canina L.	.	.	6	.	.	.	.	3	2	.	1	1	3	.	.	.	.	.	.	5	6	.	.	.	.	.	.	.	.	.	.	.	.	.	.
108. Parnassia palustris L.	.	.	.	8	.	.	.	5	.	.	.	.	.	.	.	.	.	.	.	.	.	.	.	.	.	.	.	.	.	.	.	.	.	.	.

38

Table B2. Results of ranking species based on FRANKI, option ID7 = 0.
The data set is in Table 1.

#	Species	Rank	Weight	Residual
7	Bromus erectus	1	88.452	3221.7
3	Anthoxanthum odoratum	2	83.177	3138.1
17	Festuca rubra	3	80.912	3057.2
5	Brachypodium pinnatum	4	77.914	2979.3
77	Leontodon hyspidus	5	76.458	2902.8
31	Lotus corniculatus	6	73.728	2829.1
53	Cruciata glabra	7	73.589	2755.5
83	Plantago lanceolata	8	72.777	2682.7
81	Peucedanum oreoselinum	9	72.581	2610.2
88	Potentilla erecta	10	71.680	2538.5
96	Rhinanthus freynii	11	68.818	2469.7
6	Briza media	12	68.619	2401.0
16	Festuca tenuifolia	13	65.522	2335.5
20	Koeleria pyramidata	14	65.328	2270.2
2	Agrostis tenuis	15	64.433	2205.8
36	Trifolium pratense	16	62.749	2143.0
15	Danthonia decumbens	17	62.706	2080.3
11	Carex montana	18	62.226	2018.1
13	Dactÿlis glomerata	19	57.401	1960.7
65	Galium verum	20	55.689	1905.0
4	Arrhenatherum elatius	21	55.089	1849.9
76	Knautia drymeia	22	54.955	1795.0
35	Trifolium montanum	23	51.857	1743.1
19	Holcus lanatus	24	48.162	1694.9
30	Lathyrus pratensis	25	45.684	1694.2
14	Danthonia alpina	26	42.579	1606.7
8	Crysopogon gryllus	27	42.094	1564.6
49	Leucanthemum vulgare	28	40.709	1523.9
45	Betonica officinalis	29	39.517	1484.4
23	Nardus stricta	30	38.340	1446.0
107	Viola canina	31	37.937	1408.1
44	Buphthalmum salicifolium	32	37.897	1370.2
38	Trifolium rubens	33	37.074	1333.1
18	Festuca pratensis	34	36.954	1296.1

Card 4 – Name of data set and specification NQ, ITR, IFPR

```
                        40       45      50      55
----------------------I------I-------I------I-----------------------
GRASSLANDS               35       0      1
-----------------------------------------------------------------------
```

Card 5 and following cards – data in format (35F2.0)

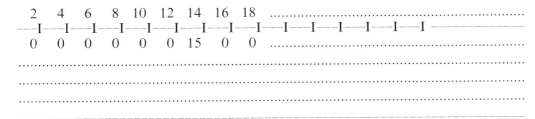

```
  2    4    6    8   10   12   14   16   18  ..................................................
--I---I---I---I---I---I---I---I---I---I---I---I---I---I--  --------------
  0    0    0    0    0    0   15    0    0  ..................................................
      ..................................................................................
      ..................................................................................
      ..................................................................................
----------------------------------------------------------------------
```

Card 6 – Left blank

```
      ----------------------------------------------------------------
      ----------------------------------------------------------------
```

Abbreviations:

NTB	=	< or >0 (number of sets to be compared; negative when sets are subdivisions of a table, positive when sets are independent tables; must be 1 for SINGLEHET, CLACOMP and DISCAN
INF	=	5 or 1, 2, 3 (input from cards or tape 1, tape 2, tape 3; always 5 or 1 for SINGLEHET, CLACOMP and DISCAN)
IFUNCT	=	INF1F, INF1, INF2, DAB, SINGLEHET, CLACOMP, or DISCAN (specifications for information function or operation, left justified)
NDPW	=	1, 2, 4 or 10 (specification for the number of data elements to be packed in one word)
NROWS	=	specification of number of rows
LOUTDIS	=	1, 2 or 3 (specification of tape for writing the triangular resemblance matrix)
NDIG	=	specification for total number of digits (right and left of decimal point)
NDEC	=	specification for number of digits right of decimal point
IND	=	tape designation for table input; if unspecified, IND = INF assumed
NQ	=	specification for number of columns; NROWS * NQ can range from 20,000 to 200,000, depending on the packing chosen (NDPW)
ITR	=	0 or >0 (option for transposition; 0 for no transposition)
IFPR	=	0 or >0 (option for printing data; 0 for no printing).

As the cards are set up in the example, a resemblance matrix is computed according to function (38) for all relevés described by the species in Table B2. Output is not given.

D. Cluster analysis

The relevés of Table B1 are clustered based on different algoritms:

Algorithm	Resemblance function
CINF1F	(38)
CINF2	(36)
CDAB	(34)

The reduced species list of Table B2 is used. The clustering criterion is (56). The results are given in Figure D1.

To run CINF1F the following cards are needed:

Card 1 – Specification for M, N, CDOL, T1, INF, IFBLA, ITR

```
      5      10     15     20     25     30     35
------  I----  I----  I----  I----  I----  I----  I ----------------------------
     34     35                5      1      0
------------------------------------------------------------------------------
```

Card 2 – Input format for data (IFMT)

```
------------------------------------------------------------------------------
(35F2.0)
------------------------------------------------------------------------------
```

Card 3 – Specification for printing data table (LFMT)

```
------------------------------------------------------------------------------
(35F3.0)
------------------------------------------------------------------------------
```

Card 4 and following cards – data if input is not from Tape 1 or 2

```
------------------------------------------------------------------------------
```

To run CINF2 or CDAB the following cards are needed:

Card 1 – Specification for M, N, INF, IFBLA, ITR

```
      5      10     15     20     25
------  I----  I----  I----  I----  I ------------------------------------------
     34     35      5      1      0
------------------------------------------------------------------------------
```

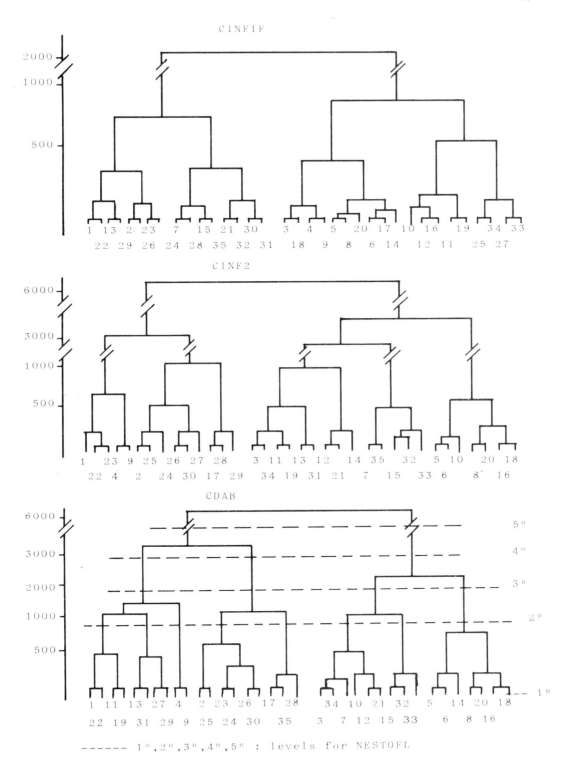

Figure D1. Dendograms of relevés from different programs. The data set is in Table A1.

42

Card 2 – Input format for data (IFMT)

(35F2.0)

Card 3 – Specification for printing data table (LFMT)

(35F3.0)

Card 4 and following cards – data if input is not from Tape 1 or 2

Abbreviations:

M   = number of rows (up to 100)
N   = number of columns (up to 100)
CDOL   = CONT or left blank (if CONT, continuous variables are categorized)
T1   = a number specifying the class interval in standard deviation units; Tl is active when CDOL
                  = CONT
INF   = 5 or 1, 2 (input from card or Tape 1, Tape 2)
IFBLA   = output file for the sequence of the classified objects produced by the program
ITR   = 0 or >0 (>0 to transpose data matrix)
IFMT   = input format for data (FORMAT (8A10))
LFMT   = format for printing data table (FORMAT 8A10)).

Classification efficiency at any level of hierarchical in CINF1F, CINF2 and CDAB can be computed. An example is given for each at the 4-cluster level:

CINF1F

Source	Information	Efficiency
Between groups	$IB = 2011 - 1419 = 592$	0.29
Within groups	$IW = 312.62 + 314.95 + 364.70 + 427.68 =$	
	$= 1419$	

CINF2

Source	Information	Efficiency
Between groups	$IB = 6333.41 - 4369.88 = 1963.53$	0.31
Within groups	$IW = 1679.80 + 1175.00 + 1852.22 +$	
	$662.86 = 4369.88$	

CDAB

Between groups	IB = 6768.42 − 2964.03 = 3804.39	0.56
Within groups	IW = 1602.63 + 1131.11 + 1094.40 +	
	722.25 = 2964.03	

Measures of efficiency are not directly comparable because they are measured in different information quantities. They serve only to compare the efficiency among levels within the same classification (see Feoli & Lausi, 1980). Options SINGLEHET and CLACOMP of SINFUN, MLTAX2 of IAHOPA and program NESTOFL perform the comparison.

E. Predictivity analysis

Option CLACOMP of SINFUN can be used to compare classifications on the basis of class contents or on the basis of their predictivity with respect to external variables in conformity with (36). The following are examples of comparison of classifications CINF1F and CINF2 of Fig. D1 in terms of class contents at the 4-cluster level and for predictivity of CINF1F and CINF2 with respect to pH. The latter analysis is analogous to that proposed by Guillerm (1971) for single species and ecological factors.

The contingency table for comparison of the class contents in CINF1F and CINF2 at four clusters and results from CLACOMP are as follows:

```
                    Clusters in CINF1F
                   1    2    3    4

            1      3    0    2    0
Clusters    2      4    2    1    2
in
CINF2       3      1    6    2    5
            4      0    0    5    2
```

$2I(A_u;A_z) = 25.185$ (9 degrees of freedom with chi square probability in left tail 0.997)

Components (17, 24, 18, 25)

Between rows	92.094
Within rows	71.546
Between columns	96.730
Within columns	66.909

Interaction shared by rows (58)

Row	$2I(U_i;Z)$	Degrees of freedom	Chi square probability $\times$ 100
4	9.5843	3	97.7
1	7.1364	3	93.2
3	5.7295	3	87.4
2	2.7344	3	56.5

Interaction shared by columns (58)

Column	$2I(U;Z_j)$	Degrees of freedom	Chi square probability $\times$ 100
1	8.7840	3	96.7
2	7.4306	3	94.0
3	5.8473	3	88.0
4	3.1227	3	62.7

The following give the pH values for the relevés in Table B1, contingency tables for predictivity analysis, and results from CLACOMP:

Relevé	1	2	3	4	5	6	7	8	9	10	11	12	13	14	15	16	17	18
pH $\times$ 10	64	61	55	64	53	52	61	51	56	53	62	57	61	52	63	53	75	53

	19	20	21	22	23	24	25	26	27	28	29	30	31	32	33	34	35
	59	55	59	60	61	75	67	77	77	69	63	64	58	58	65	64	69

		Clusters in CINF1F			
		1	2	3	4
	1	0	0	9	3
Classes	2	1	3	0	1
of	3	4	3	0	3
pH	4	0	1	0	1
	5	2	0	1	1

$2I(A;A_z) = 34.815$ (12 degrees of freedom with chi square probability 0.99 in left tail)

Components (17, 24, 18, 25):

Between rows	95.123
Within rows	55.867
Between columns	90.682
Within columns	60.308

Interaction shared by rows (58)

Row	$2I(U_i;Z)$	Degrees of freedom	Chi square probability $\times$ 100
1	15.790	3	99.8
3	7.726	3	94.8
2	5.501	3	86.1
4	2.927	3	59.7
5	2.871	3	58.8

Interaction shared by columns (58)

Column	$2I(U;Z_j)$	Degrees of freedom	Chi square probability $\times$ 100
3	15.928	4	99.7
2	10.033	4	96.0
1	8.387	4	92.2
4	.468	4	2.3

		Clusters in CINF2			
		1	2	3	4
	1	0	0	2	7
Classes	2	2	0	5	0
of	3	3	3	6	0
pH	4	0	2	1	0
	5	0	4	0	0

$2I(A_u;A_z) = 45.411$ (12 degrees of freedom with chi squared probability 1.00 in left tail)

Components (17, 24, 18, 25)

Between rows	104.760
Within rows	46.683
Between rows	92.094
Within columns	59.351

Interaction shared by rows (58)

Row	$2I(U_i;Z)$	Degrees of freedom	Chi squared probability $\times$ 100
1	16.663	3	99.9
5	10.865	3	98.7
2	8.570	3	96.4
3	5.866	3	88.1
4	3.446	3	67.2

Interaction shared by columns (58)

Column	$2I(U;Z_j)$	Degrees of freedom	Chi squared probability $\times$ 100
4	19.014	4	99.9
2	14.507	4	99.4
1	6.130	4	81.0
3	5.760	4	78.2

To run SINFUN with option CLACOMP the following cards are needed:

Card 1 – Title and specifications NTAB, INF, IFUNCT, NDPW and NROWS

```
              30   35   40        50   55   60   65   70
               I    I    I         I    I    I    I    I
CINF1F-CINF2        1    5 CLACOMP      1    4    0
```

Card 2 – Input format for data and specification NDIG, NDEC

```
                                   70   75   80
                                    I    I    I
(4F1.0)                                  1    0
```

Card 3 – Format specification for printing data table

```
(1X,4F5.2)
```

Card 4 – Title and specifications NQ, ITR, IFPR

```
              40   45   50   55
               I    I    I    I
PREDICTIVITY        4    0    1
```

Card 5 and following card – data in format (4F1.0)

Card 6 – Blank card

Predictivity can be measured based on (34) by option SINGLEHET of SINFUN. To run SINFUN with option SINGLEHET for contingency tables (pH and classes CINF1F; pH and classes CINF2) the following cards are needed:

Card 1 – Title and specifications NTB, INF, IFUNCT, NDPW, NROWS and LOUTDIS

```
0            30    35    40    45    50    55    60    65    70
----------I----I----I----I----I----I----I----I----I----------
SINGLEHET      1     5 SINGLEHET    1     5     0     0
```

Cards 2, 3, 4, 5, 6 as for CLACOMP.

Abbreviations have already been explained in C. The heterogeneity value given by SINGLEHET for the contingency table of pH and CINF1F is 37.354. It is 50.358 for pH and CINF2. From these results and those of CLACOMP we can conclude that the classification given by CINF2 is more predictive for pH than is the classification given by CINF1F.

F. Nested model

The nested model in Table 4 could be applied in a spatial pattern analysis (Orlóci, 1971) or in an analysis of the discrimination power of species at different hierarchical levels of a classification. An example of computation has been already given in Table 5. In the present table an output of program NESTOFL is presented for species in Table B2 (coded according to rank order). Five hierarchic levels of classification from CDAB (Fig. D1) are considered. The columns of the output give the information, species ID number and chi square probability ($\times$ 100):

SORTED LEVEL N.   1

INF.VALUE-	SPECIES N.-	X2PROBAB.
104.59	11	100.0
84.842	5	100.0
84.601	18	100.0
78.483	3	100.0
76.758	13	100.0
75.741	2	100.0
70.710	15	100.0
68.239	21	100.0
58.347	26	100.0
54.699	9	99.9
54.693	24	99.9

54.138	1	99.9
53.282	25	99.8
48.912	16	99.4
47.503	20	99.1
47.105	17	99.0
46.503	7	98.9
44.847	34	98.3
43.191	10	97.5
43.099	14	97.4
42.000	6	96.7
40.540	22	95.4
40.389	30	95.3
37.475	8	91.3
37.309	28	91.1
34.272	19	84.2
29.538	29	66.5
28.150	31	59.7
25.547	4	45.6
24.140	32	37.7
22.939	12	31.2
22.325	23	27.9
21.344	27	23.0
19.883	33	16.4

Sum of $X2$ PROBAB. = 2851.1

SORTED LEVEL N.   2

INF.VALUE- SPECIES N.- X2PROBAB.

36.865	7	100.0
45.366	10	100.0
45.694	15	100.0
32.560	24	100.0
28.402	11	100.0
27.563	1	100.0
26.380	13	100.0
24.032	22	100.0
23.870	3	100.0
23.789	9	100.0
22.389	12	100.0
21.525	21	100.0
21.522	18	100.0
20.702	34	100.0
20.609	4	100.0
18.685	28	99.9
17.477	14	99.8

17.466	16	99.8
16.462	17	99.8
15.392	23	99.6
12.438	8	98.6
12.325	27	98.5
10.310	31	96.4
10.270	20	96.4
9.7793	26	95.6
9.2470	25	94.5
8.6880	19	93.1
7.6227	29	89.4
7.4802	2	88.7
6.2657	5	82.0
6.0949	33	80.8
4.9508	6	70.8
1.3244	32	14.3
.13278E-01	30	.0

Sum of X2 PROBAB. = 3098.0

SORTED LEVEL N.   3

INF.VALUE-	SPECIES N.-	X2PROBAB.
50.057	1	100.0
22.469	5	100.0
18.040	7	100.0
43.150	11	100.0
20.301	18	100.0
17.505	30	100.0
10.932	23	99.9
10.845	8	99.9
10.329	2	99.9
9.7812	27	99.8
6.9999	29	99.2
6.9754	14	99.2
6.4644	12	98.9
5.7536	20	98.4
4.7750	10	97.1
4.7040	28	97.0
4.3566	3	96.3
4.0275	33	95.5
3.7137	32	94.6
2.8768	34	91.0
2.6571	9	89.7
2.1613	4	85.9
1.8226	17	82.3

1.8127	13	82.2
1.7261	19	81.1
1.6534	21	80.2
1.5349	24	78.5
1.5147	15	78.2
1.1807	6	72.3
1.1152	22	70.9
.99480	26	68.2
.54291	31	53.9
.26571	25	39.4
.13807	16	29.0

Sum of X2 **PROBAB.** = 2958.5

SORTED LEVEL N.   4

INF.VALUE-	SPECIES N.-	X2PROBAB.
43.251	2	100.0
37.288	9	100.0
35.859	10	100.0
17.095	15	100.0
17.882	20	100.0
16.439	32	100.0
19.621	34	100.0
12.669	27	100.0
12.572	24	100.0
11.130	13	99.9
11.008	16	99.9
10.270	31	99.9
8.8980	5	99.7
8.5404	18	99.7
7.9610	11	99.5
7.3203	25	99.3
6.1464	8	98.7
6.0125	17	98.6
5.6819	12	98.3
5.0119	7	97.5
3.3875	23	93.4
3.1851	21	92.6
3.1299	14	92.3
2.9889	26	91.6
2.0556	29	84.8
1.5322	1	78.4
1.0268	6	68.9
.32459	19	43.1
.23819	4	37.5

.21965	22	36.1	
.10569	33	25.5	
.84317E-01	28	22.8	
.22881E-02	3	3.8	
0.	30	0.0	Sum of X2 PROBAB. = 2761.8

SORTED LEVEL N.   5

INF.VALUE- SPECIES N.- X2PROBAB.

21.494	7	100.0
20.395	10	100.0
38.145	13	100.0
15.212	16	100.0
47.556	17	100.0
26.992	18	100.0
37.741	19	100.0
20.928	20	100.0
73.003	21	100.0
21.174	24	100.0
19.731	25	100.0
47.516	26	100.0
62.621	30	100.0
20.828	34	100.0
13.528	4	100.0
10.371	29	99.9
8.7662	6	99.7
8.2649	15	99.6
3.9572	9	95.3
2.8304	8	90.8
2.2820	11	86.9
2.2406	28	86.6
1.6403	31	80.0
1.6156	27	79.6
1.3775	33	76.0
1.0570	12	69.6
.75867	32	61.6
.74630	23	61.2
.72673	5	60.6
.56754	14	54.9
.15244	22	30.4
.56752E-01	1	19.0
.20795E-01	3	11.5
.53132E-02	2	5.8

Sum of X2 PROBAB. = 2769.0

NESTOFL TERMINATED

The sum of probabilities, given by the program, can help to define the optimal hierarchical level.

52

To run NESTOFL the following cards are needed:

Card 1 – Title and specifications NESTYP1 or NESTYP2

```
                              40        50        60
--------------------------------I---------I---------I-----------------
NESTING BY CDAB       HIERARCHIC BINARYTWOI
```

Card 2 – Specifications NSPEC, NQUAD, ITRASP, IFPR and INF

```
     5    10    15    20    25    30    35
-----I-----I-----I-----I-----I-----I-----I---------------------
   34    35     0     0     5
```

Card 3 – Input format for data

```
(35F2.0)
```

Card 4 – Specification of format for printing the data

```
(1X,35F3.0)
```

Card 5 and following cards – data in format (35F2.0)

Card 6 – Specification NLEV

```
     5
-----I---------------------------------------------------
   5
```

Card 7a – Specification NGR at the second level

```
     5
-----I---------------------------------------------------
   8
```

Card 8a – Specification of column subscript for the first element in each block at the second level

```
     5    10    15    20    25    30    35    40    45
-----I-----I-----I-----I-----I-----I-----I-----I-----I---------
   5     9    11    17    20    26    29
```

Card 7b – Specification NGR at third level

```
   5
----I --------------------------------------------------------
   4
```

Card 8b – Specification of column subscripts for the first element of each block at third level

```
   5    10    15    20    25    30
----I----I----I----I----I----I --------------------------------
  11    20    29
```

Card 7c – Specification NGR at fourth level

```
   5
----I --------------------------------------------------------
   3
```

Card 8c – Specification of column subscript for the first element of each block at fourth level

```
   5    10    15    20    25
----I----I----I----I----I --------------------------------------
  11    20
```

Card 7d – Specifications NGR at fifth level

```
   5
----I --------------------------------------------------------
   2
```

Card 8d – Specification of column subscript for the first element of each block at fifth level

```
   5
----I----I --------------------------------------------------
  20
```

Abbreviations:

NESTYP1	=	HIERARCHIC (the nested model in Table 4 performed)
NESTYP2	=	BINARYTWOI (Interaction according to (36) computed at each level between each species and the classes)
NSPEC	=	number of species (up to 1000)
NQUAD	=	number of relevés
ITRASP	=	0 or >0 (>0 to transpose data table)
IFPR	=	0 or >0 (>0 to print data)
INF	=	5 or 1, 2 (cards, tape 1 or tape 2 for mass storage)
NLEV	=	number of hierarchical levels (note that row data constitute first level in NLEV)
NGR	=	number of blocks at each hierarchical level.

G. Identification

An example of identification is given based on 9 relevés to be assigned to the four main clusters of CDAB (Fig. D1). The 9 relevés are listed in the following table:

Species	Relevé 1	2	3	4	5	6	7	8	9
1	8	10	6	1	6	0	0	10	15
2	7	15	0	12	3	16	6	25	0
3	3	1	22	4	0	4	0	0	0
4	8	0	0	4	2	4	0	0	0
5	10	0	3	0	10	25	12	8	6
6	2	9	0	0	9	0	0	5	3
7	0	5	5	0	18	0	0	0	0
8	10	0	6	0	4	15	4	6	4
9	0	11	0	1	12	0	0	0	0
10	0	18	8	5	0	0	0	0	3
11	4	0	18	18	2	0	0	20	25
12	18	3	6	2	5	0	0	4	0
13	0	12	5	22	3	0	0	0	0
14	2	4	0	10	0	0	0	0	0
15	0	0	2	0	0	0	0	0	0
16	0	2	0	2	0	12	6	5	0
17	0	0	8	25	0	0	0	0	0
18	8	23	16	0	0	0	0	0	15
19	7	1	0	0	0	0	10	2	1
20	0	1	0	0	9	0	0	0	0
21	1	0	0	0	3	22	23	20	0
22	3	1	0	6	4	5	15	1	2
23	22	3	0	0	0	0	0	0	5
24	0	0	0	0	0	8	0	22	0
25	0	0	0	0	0	6	0	2	3
26	4	0	0	2	0	0	0	0	0
27	0	0	8	0	0	0	0	0	0
28	3	3	0	0	5	0	2	6	4
29	0	1	3	4	0	0	0	0	0
30	0	0	0	5	0	0	0	0	0
31	0	4	5	0	0	0	5	0	3
32	4	0	0	0	15	0	2	0	0
33	0	0	0	0	4	0	0	0	0
34	2	0	0	0	0	0	0	7	0

Species are in the same order as in Table B2.

The results of DISCAN are the following:

Relevé to be assigned	Reference syntaxon of assignment	Nearest relevé in reference syntaxon
1	3	28
2	3	20
3	3	23
4	4	33
5	2	17
6	2	13
7	1	9
8	1	9
9	3	28

Total sample heterogeneity = 1867.1
Total within taxa heterogeneity = 888.12
Efficiency (between taxa/total sample) = 0.52.

To run SINFUN with option DISCAN the following cards are needed:

Card 1 – Title and specifications as in Table 9

```
                        30I--35I--40I----------50I--55I--60I
Identification           1     5 DISCAN        1    34
```

Card 2 – Input format and specification NDIG, NDEC

```
                        70I--75I--80I
(9F2.0)                  2    0
```

Card 3 – Format specification for printing data table (left blank for no printing)

```
                70I
(1X,9F3.0)
```

Card 4 – Title and specifications NQ, ITR, IFPR

```
                        40I--45I--50I--55|
GRASSLANDS IDENT.        9    0    0|
```

Card 5 and following cards – data for N objects to be assigned in format (9F2.0)

Card 6 – Specifications NTAXA, NOBTAX, LOFMT

```
    5        10
-----I------I------------------------------------------------------
    4    35   (I2,35F2.0)
----------------------------------------------------------------------
```

Card 7 and following cards – data for reference taxa with specified NTAXON in format LOFMT.

Abbreviations for DISCAN:

NDIG	=	⎫
NDEC	=	⎬
NQ	=	⎬ Defined in program SINFUN
ITR	=	⎬
IFPR	=	⎭
NTAXA	=	2 or >2 specifying the number of reference taxa (no upper limit)
NOBTAX	=	1 or >1 specifying the number of objects in reference taxa (no upper limit)
LOFMT	=	input format of objects in reference taxa
NTAXON	=	numerical label of the taxon of objects in reference taxa.

H. Structuring data tables

This accords with the sequence of species based on the nearest neighbour criterion applied to a matrix of species associations computed according to Rajski's metric (22) for diversity arrays. Comparison of sets in different tables or subsets in a single table is according to (53). Program IAHOPA performs the computations. IAHOPA is an overlay program with four branches. A simple root program MASTER allows to select the desired branch. MASTER requires two cards:

Card 1 – Title for IAHOPA run

```
                        40
------------------------I------------------------------------------
IAHOPA TESTS
----------------------------------------------------------------------
```

Card 2 – Calling card for the branches with specification for MLTAX, INFILE, IFCUT, LIN, LOUT, LET

```
        10    15    20    25    30    35
--------I-----I-----I-----I-----I-----I----------------------------
MLTAX1      5     1     3     3     1
----------------------------------------------------------------------
```

or

	10	15	20	25	30	35	
SCRAPS	3	0	0	3	4		

or

	10	15	20	25	30	35
MLTAX2	3	0	0	0	0	

or

	10	15	20	25	30	35
MLTAX3	3	0	0	0	0	

Abbreviations:

MLTAX	=	name of the branch (MLTAX1, SCRAPS, MLTAX2, MLTAX3)
INFILE	=	5 or 8 (input from cards or tape 8)
IFCUT	=	0 or >0 (if >0 MLTAX1 writes file LIN, the data for SCRAPS)
LIN	=	0 or 3 (if IFCUT >0, LIN must be 3, otherwise 0)
LOUT	=	0 or 3 (if 3, SCRAPS writes results on tape 3, 0 for no writing)
LET	=	number of subsets to be created by SCRAPS from the data table.

If MLTAX1 is selected the following are the IAHOPA cards:

Card 3 – Title card for run MLTAX1

	40	
GRASSLANDS		

Card 4 – Specifications NSPEC, NQUAD, NUDATA, INTRASP, IFPR, LASOLP, INSEQ, ISIM, IFULL, IRAJSKI

5	10	15	20	25	30	35	40	45	50
34	35	35	0	0	1	0	0	0	1

Card 5 and following cards – data for species codes and names (ISPEC, ILCOPI, NOUN)

```
 3          10
--I---------I ---------------------------------------------------------------
 1 0000001    Helictotrichon pratense
 ...........................................................................................................................
 ...........................................................................................................................
 ...........................................................................................................................
 ...........................................................................................................................
--------------------------------------------------------------------------------
```

Card 6 – Input format for data table (IF)

```
--------------------------------------------------------------------------------
(35F1.0)
--------------------------------------------------------------------------------
```

Card 7 and following cards – data in format (IF).

If the user wants a print of the structured table the data must have only one digit numbers. If the data have more than one digits MLTAX1 gives the sequence of species but the structured table is printed with random symbols. A transformation to 1 digit numbers is necessary to run MLTAX1 as in the following example where the data of Table B1 have been transformed. Only the 34 species of Table B2 are considered.

Last card – branch termination card

```
               10
---------------I -----------------------------------------------------------
FINEMLTAX1
--------------------------------------------------------------------------------
```

Abbreviations:

NSPEC	=	number of species (up to 118)
NQUAD	=	number of relevés (up to 1000 if ITRASP >0; otherwise up to 250)
NUDATA	=	number of data in an input row (if ITRASP $= 0$ NUDATA must be equal to NQUAD)
ITRASP	=	0 or >0 (if 0, data entry by species, if >0 data entry by relevés)
IFPR	=	0 or >0 (if >0, similarity matrix printed)
LASOLP	=	0 or >0 (if >0, binary transformation of data)
INSEQ	=	0 or n (if n, data table structured n times according to n external sequences of species, stored in vector LORDER with format (16 I5); no similarity matrix computed; LORDER is the last set of data cards before last card)
ISIM	=	0 or >0 (if >0, similarity matrix written on tape 4, available for clustering or ordering algorithms)
IFULL	=	0 or >0 (if >0, relevés ordered according to quantitative scores; if $= 0$, relevés ordered by presence/absence scores)

IRAJSKI   = 0 or >0 (if >0, Rajski's metric computed, otherwise similarity ratio)
ISPEC   = position column index of species to allow selection of the relevant species from NUDATA
ILCOPI   = numerical code for species (any number up to 7 digits)
NOUN   = species names.

The dendrogram table produced by MLTAX1 for the 34 species of Table B2 is the following (the code for species is the rank in Table B2):

DENDROGRAM TABLE			1	.9	.8	.7	.
1	33	.97285	1	33.			
1	12	.96675	1	12			
1	32	.96625	1	32			
1	23	.96609	1	23			
1	14	.96477	1	14			
1	8	.96223	1	8			
1	27	.95814	1	27			
1	4	.95792	1	4			
27	3	.95705	27	3			
33	7	.95685	33	7.			
4	22	.95475	4	22			
1	19	.95464	1	19			
22	2	.95416	22	2			
2	31	.96911	2	31			
2	29	.96217	2	29			
29	10	.96523	29	10			
1	6	.95413	1	6			
2	16	.95273	2	16			
1	28	.95231	1	28			
1	5	.95202	1	5.			
10	9	.94901	10	9			
10	13	.94740	10	13			
13	17	.96780	13	17			
7	21	.94735	7	21			
21	25	.94855	21	25			
2	34	.94586	2	34			
1	18	.94426	1	18			
1	20	.94413	1	20			
10	15	.94058	10	15			
17	30	.93996	17	30.			
7	24	.93964	7	24			
5	11	.93506	5	11			
1	26	.92585	1	26			

The dendogram table allows to build an MST graph (Minimum Spanning Tree). By decomposing the MST according to the maximal Rajski's metric, groups of species could be obtained. The data table (original data have been grouped into 9 classes of frequency) is structured into 6 subtables by the sequence of species given by the program:

SPECIES

```
31321 2   21 321 12   112231213212
13223487437292190668593715480506416
```

```
1111111111111111111111111111111111
```

```
31321 2   21 321 12   112231213212
13223487437292190668593715480506416
```

RELEVÉS

7	611.11311331....2...1..4.......23	
31	212.114.113.14.13.1.15..1..311.58.	
22	211.1.41124.4...513.17..72..18.7..	
21	51.1111.122..52.51..6.42...7....9.	I°
25	41.12.3141.332.1.3..6.......4...4.	
23	21....11336.11....1.1...331.33.4..	
24	72....1.324131...1......42..9.....	

32	9.11141.4.14.4.......42...41.2.4.	
28	3.21.1.233..4....4.14............	II°
35	9.21.1.213.......1..1.......1....1	
12	3.12..1.1.21.1..41..4112...4.23161	

11	6.3.113.3151151.3223136411122..42.	
19	1.4.126.1443.5.1512.636221...3.2..	
29	4.3.443.26..47...4..34....54......	
33	9.3.343.12...3...2115..1...8....84	III°
15	6.3.112.1..1.....2.3..63...61....6	
13	1.2.1....33..21..3..21..3..5...23.	

3	2.2..1311131.2114.1.4112...4.2....	
27	5.1..32.153.16...24261..888.....3.	
18	2.1..11..8.1.8..6.1...34...2.45.21	
34	5.1..2.1121114.1232.5.12..14.1..11	IV°
30	1.1...1144111....22..1.11..3......	
10	3.1...33162..1226...3563...5112364	
16	1.1...2.16.3.9.2312.232.......2...	
1	7.1...1.2.72271.413.78..13.4.9.4..	

26	5..2.21.41512....32.....9.1.......	
2	4..2..1222.122...2112.1.1..234.35.	V°
20	1....1..15...4212....274......4...	

17	..11.111.2...2..12.2.1......1.....	
9	.....1..8.31.9126.2.111.711..5..2.	
14	.....1.........171..1.3511.6.17...	
4	......1.4..4.91..1814...1.3..4..7.	VI°
5	........11...4.26...1351...2.11..5	
8	........1....7114....257.....33...	
6	........2....3.321...1.........1.4	

The table produced by MLTAX1 can be divided according to different criteria. Program SCRAPS allows to subdivide such a table and to prepare the data for MLTAX2 and/or MLTAX3. To subdivide the table into six subtables marked by roman numerals, SCRAPS has to be run and the following IAHOPA cards are needed:

Card 3 – Specification for number of subsets

```
----5-I --------------------------------------------------------------
    6
```

Card 4 and following cards – Specifications for bounds of subsets (row and column subscripts for left upper and right lower cell in subset).

```
   5-I--10-I--15-I--20-I--25-I--30-I--35-I--40-I--45-I--50-I--55-I--60-I--65-I--70-I--75-I--80
   1    1    7   34    8    1   11   34   12    1   17   34   18    1   25   34
```

```
   5-I--10-I--15-I--20-I--25-I--30-I--35-I--40-I--45-I--50-I--55-I--60-I--65-I--70-I--75-I--80
  26    1   28   34   29    1   35   34
```

Card 5 – Branch termination card

```
------------------10-I----------------------------------------------
FINESCRAPS
```

If program MLTAX2 is selected to measure homogeneity within the subsets given by SCRAPS, the following IAHOPA card is expected:

Card 3 – Branch termination card

```
                     10
--------------------I --------------------------------------------------
FINEMLTAX2 --------------------------------------------------------------
```

If the subsets prepared by SCRAPS are to be compared by MLTAX3 the following IAHOPA cards are needed:

Card 3 – Title for MLTAX3 run and specifications IFPR, IENT, NPAIR, IFRE, INOTAB and NEWDIAG.

```
                       40   45   50   55   60   65   70
----------------------I ---I ---I ---I ---I ---I ---I --------
MLTAX3 COMPARISONS      1    1    0    0    0    0
```

Card 4 – Branch termination card

```
                     10
--------------------I --------------------------------------------------
FINEMLTAX3 --------------------------------------------------------------
```

Abbreviations:

IFPR	= 0 or >0 (if >0, fusion matrices written)
IENT, IFRE	= 0 or >0 (operate joinly. If IENT = 0 and IFRE = 0, subsets merged; (36) and (53) computed on the co-occurrence matrix of species. If IENT >0 and IFRE = 0, subsets merged; (36) and (53) computed on data of the merged set. If IENT = 0 and IFRE >0, marginal dispersion diversity arrays are computed for species and compared by (53) in each subset.)
NPAIR	= 0 or >0 (if >0, MLTAX3 compare specified pairs of data tables. If subsets of the same table have to be compared, SCRAPS must be used. In this case the last set of data cards, before the branch termination card, should specify the pairs of sets or subsets to be compared (format 2I5). Each set or subset is automatically coded according to position in the data set)
INOTAB	= 0 or >0 (>0 for printing data in all sets or subsets)
NEWDIAG	= 0 or >0 (>0 to replace diagonal elements of the co-occurrence matrix by the frequency of species in the sample).

In the example MLTAX3 gives the following redundancy matrix to which a clustering algorithm can be applied,

	1	2	3	4	5
2	0.811				
3	0.832	0.829			
4	0.825	0.806	0.841		
5	0.808	0.793	0.838	0.811	
6	0.761	0.768	0.789	0.802	0.757

If the subsets stored in separated tables have to be compared, the cards after the third card of MLTAX3 have to be prepared as follows for each table:

Card 4 – Title of table (as in MLTAX1)

Card 5 – Specifications for NSPEC, NQUAD, ITRASP (specified in MLTAX1) and IFPR, IENT and NEWDIAG (specified in MLTAX3)

```
----5-I----10-I----15-I----20-I----25-I---------------------------------
-----------------------------------------------------------------------
```

Card 6 and following cards – same as cards 5, 6 and 7 of MLTAX1

Card 7 – Sample termination card

```
----------------I-------------------------------------------------------
FINESAMPLE-------------------------------------------------------------
```

This card can be followed by the sets of cards of another group of tables, each table with cards 4, 5, 6.

Card 8 – Branch termination card

```
--------------10-I-----------------------------------------------------
FINEMLTAX3------------------------------------------------------------
```

If only homogeneity have to be computed within the separated tables MLTAX2 requires the following cards after the first two of IAHOPA:

Card 3 – Title card for the table

Card 4 – Specifications NSPEC, NQUAD, NUDATA, ITRASP and IFPR as in MLTAX1

Card 5 and following cards similar to MLTAX1.

At the end of the data of the set of tables an end record card is needed as in MLTAX3:

64

Card 6 – Branch termination card

```
_____10_I _____
FINEMLTAX2 _____
```

Program listings are on the following pages:

7. Program listings

```
CFRANKI
      PROGRAM FRANKI(INPUT=100,OUTPUT=100,TAPE5=INPUT,TAPE1=514,TAPE2=
     1514)
C    LABEL=TITLE OF DATA TABLE(UP TO 40 CHARACTERS)
C    NP,NQ=NUMBER OF SPECIES AND OF RELEVES RESPECTIVELY.
C        NP UP TO 110,NQ UP TO 125
C    OPTIONS:
C  ID7=IF 0,RANKING ON INTERACTION INFORMATION(ORLOCI 1976)
C      =IF 1,RANKING OF EQUIVOCATION DIVERGENCE(ORLOCI 1978)
C      =IF 2,RANKING ON RAJSKI*S METRIC
C  INF=INPUT FILE FOR DATA(IF 5 ON CARDS IF 1 ON TAPE1)
C  ITR=IF 0,DATA ENTERED BY SPECIES
C      =IF 1,BY RELEVES.DATA ALWAYS STORED BY SPECIES
C  IPR=IF 1,DATA TABLE PRINTED
C  INPFMT=INPUT FORMAT FOR DATA TABLE
C  ...........................................................
C  A BLANK CARD TERMINATES FRANKI
C  ...........................................................
      COMMON X(110,125),D(125),AI(110),G(110),XK(110,125),F(125),Z(110)
      DIMENSION LABEL(4),INPFMT(8)
      IBLANK=10H
      REWIND1
1000  FORMAT(1HT)
4000  READ(5,1) (LABEL(I),I=1,4),NP,NQ,ID7,INF    ,ITR,IPR
   1  FORMAT(4A10,8I5)
      IF(LABEL(1)-IBLANK) 3000,2000,3000
3000  PRINT 2,(LABEL(I),I=1,4),NP,NQ,ID7,INF    ,ITR,IPR
   2  FORMAT(1H1,1X,4A10/19H NUMBER OF SPECIES=,I5/20H NUMBER OF QUADRAT
     1S=,I5/18H WANTED OPTION D7=,I5/17H DATA INPUT FILE=,I5/15H TRASP.
     2OPTION=,I5/14H PRINT OPTION=,I5)
      READ(5,3) (INPFMT(I),I=1,8)
   3  FORMAT(8A10)
      REWIND2
      IF(ITR-0) 4,4,8
   4  DO 5 J=1,NP
      READ(INF,INPFMT) (X(J,K),K=1,NQ)
      IF(IPR-0) 5,5,6
   6  PRINT 7,(X(J,K),K=1,NQ)
   5  CONTINUE
   7  FORMAT(1X,10G12.5)
      GO TO 11
   8  DO 9 K=1,NQ
      READ(INF,INPFMT) (X(J,K),J=1,NP)
```

```
       IF(IPR-0) 9,9,10
 10    PRINT 7,(X(J,K),J=1,NP)
  9    CONTINUE
 11    DO 12 J=1,NP
 12    WRITE(2)(X(J,K),K=1,NQ)
       NP1=NP
            IR=1
       CALL GOTI(NP,NQ)
 43    CALL UNNI(NP1,TI,S,NQ)
       T=0.
            ZS=1.
       DO 13 JH=1,NP
       IF(AI(JH)+111111.) 14,13,14
 14    T=T+AI(JH)
       ZS=ZS*Z(JH)
 13    CONTINUE
       T1=T-TI
       IF(ID7) 15,17,15
 15    FNQ=FLOAT(NQ)
       TI1=S-FNQ*ALOG(FNQ/ZS)
       T1=TI1-T1
       IF(ID7-2) 150,160,160
160    T1=T1/TI1
150    PRINT 16,NP1,T1
 16    FORMAT(22HOEQU.INF.IN REMAINING ,I5,8HSPECIES ,G12.5)
       GO TO 19
 17    PRINT 18,NP1,T1
 18    FORMAT(25HOMUTUAL INF.OF REMAINING ,I5,8HSPECIES ,G12.5)
 19    L1=0
       NP1=NP1-1
 28    L1=L1+1
       IA=ID=0
       ZS=1.
            REWIND2
         T=0.
       DO 20 JH=1,NP
       READ(2)(D(J),J=1,NQ)
       IF(AI(JH)+111111.) 21,20,21
 21    ID=ID+1
       IF(L1-ID) 22,20,22
 22    IA=IA+1
       DO 23 J=1,NQ
 23    X(IA,J)=D(J)
       T=T+AI(JH)
       ZS=ZS*Z(JH)
 20    CONTINUE
```

```
      CALL UNNI(NP1,TI,S,NQ)
      IF(ID7) 25,24,25
  25  TI1=S-FNQ*ALOG(FNQ/ZS)
      G(L1)=T1-(TI1-(T-TI))
      IF(ID7-2) 26,260,260
 260  G(L1)=T1-(TI1-(T-TI))/TI1
      GO TO 26
  24  G(L1)=T1-(T-TI)
  26  IF(L1-(NP1+1)) 28,27,27
  27  IA=0
      GM=-1.
      DO 29 JH=1,NP
      IF(AI(JH)+111111.) 30,29,30
  30  IA=IA+1
      IF(GM-G(IA)) 31,29,29
  31  GM=G(IA)
      IS=JH
          IS1=IA
  29  CONTINUE
      IA=0
      IF(IR-NP) 33,32,33
  33  IF(ID7) 35,34,35
  35  PRINT 36,IS,IR,G(IS1)
  36  FORMAT(9HOSPECIES ,I5,10H HAS RANK ,I5,13H AND EQU.INF.,G12.5)
      GO TO 39
  34  PRINT 37,IS,IR,G(IS1)
  37  FORMAT(9HOSPECIES ,I5,10H HAS RANK ,I5,16H AND MUTUAL INF.,G12.5)
      GO TO 39
  32  PRINT 38,IS,IR
  38  FORMAT(9HOSPECIES ,I5,10H HAS RANK ,I5)
      GO TO 4000
  39  IR=IR+1
      AI(IS)=-111111.
      REWIND2
      DO 40 JH=1,NP
      READ(2)(D(J),J=1,NQ)
      IF(AI(JH)+111111.) 41,40,41
  41  IA=IA+1
      DO 42 J=1,NQ
  42  X(IA,J)=D(J)
  40  CONTINUE
      GO TO 43
2000  PRINT 5000
5000  FORMAT(18HOFRANKI TERMINATED)
      STOP
      END
```

```
      SUBROUTINE GOTI(NP,NQ)
      COMMON X(110,125),D(125),AI(110),G(110),XK(110,125),F(125),Z(110)
      DO 2 JH=1,NP
      DO 1 JM=1,125
   1  F(JM)=1.
      XK(JH,1)=X(JH,1)
      IQ=1
      DO 3 J=2,NQ
      A=X(JH,J)
      DO 4 L=1,IQ
      IF(XK(JH,L)-A) 4,40,4
  40  F(L)=F(L)+1.
      GO TO 3
   4  CONTINUE
      IQ=IQ+1
      XK(JH,IQ)=A
   3  CONTINUE
      S=0.
      DO 5 J=1,IQ
      A=F(J)
   5  S=S+A*ALOG(A)
      FNQ=FLOAT(NQ)
      AI(JH)=FNQ*ALOG(FNQ)-S
   2  Z(JH)=IQ
      RETURN
      END
      SUBROUTINE UNNI(NP1,TI,S,NQ)
      COMMON X(110,125),D(125),AI(110),G(110),XK(110,125),F(125),Z(110)
      DO 1 JH=1,NP1
   1  XK(JH,1)=X(JH,1)
      DO 2 J=1,125
   2  F(J)=1.
      IQ=1
      DO 3 J=2,NQ
      DO 4 L=1,IQ
      DO 5 JH=1,NP1
      A=X(JH,J)
      IF(XK(JH,L)-A) 4,5,4
   5  CONTINUE
      F(L)=F(L)+1.
      GO TO 3
   4  CONTINUE
      IQ=IQ+1
      DO 6 JH=1,NP1
   6  XK(JH,IQ)=X(JH,J)
   3  CONTINUE
```

```
      S=0.
      DO 7 L=1,IQ
      A=F(L)
    7 S=S+A*ALOG(A)
      FNQ=FLOAT(NQ)
      TI=FNQ*ALOG(FNQ)-S
      RETURN
      END
CSINFUN
        PROGRAM SINFUN(INPUT=100,OUTPUT=100,PUNCH=100,TAPE5=INPUT,TAPE6=OU
     1TPUT,TAPE7=PUNCH,TAPE1=514,TAPE2=514,TAPE3=514)
C--------FILENAMES,PARAMETERS IN SINFUN:
C   IDATA=DATA TABLE(MAY CONTAIN FROM 20000 TO 200000 NUMBERS,
C          DEPENDING ON THE PACKING REQUESTED)
C   NTB=NUMBER OF DATA SETS
C       IF GT.0,DATA SETS ARE INDEPENDENT TABLES.SPECIES
C       MUST BE IN IDENTICAL ORDER
C       IF LT.0, DATA SETS ARE SUBSETS OF A TABLE
C       IF LT.0 AND EQ. TO NCOL DATA SETS ARE COLUMN VECTORS.IT MUST
C       BE EQ.1,FOR OPTIONS SINGLEHET,CLACOMP,DISCAN
C   NROWS=NUMBER OF ROWS(CHARACTERS)IN TABLES. IT MUST BE SAME IN
C          ALL TABLES TO BE COMPARED
C   NDPW=NUMBER OF DATA PER WORD(1,2,4,10)
C   INF=CODE(5 OR 1,2,3) FOR INPUT FILE FOR DATA(CARDS OR TAPE1,
C       TAPE2,OR TAPE3).NO TAPE2 OR 3 ALLOWED WHEN  IFUNCT=DISCAN
C   IFUNCT= DEFINED FUNCTION DEPENDING TO SUBROUTINE:
C           =INF1
C           =INF1F
C           =INF2
C           =INF3
C           =INF3F
C           =DAB
C           =SINGLEHET
C           =CLACOMP
C           =DISCAN
C   LOUTDIS=NAMES OF FILE INTO WHICH D(J,K) OR D(A,B) ARE WRITTEN
C          (FORMAT5X,10G12.5)
C--------SUBROUTINE STRUCOL:
C   IFM=INPUT FORMAT FOR DATA(70 CHARACTERS)
C   NDIG=NUMBER OF DIGITS IN A DATUM
C   NDEC=DIGITS TO THE RIGHT OF DECIMAL POINT,IF ANY
C   LFM=LIST FORMAT FOR DATA(70 CHARACTERS)
C   IND=INPUT FILE FOR DATA SETS.IF NOT SPECIFIED,IND=INF
C   FROM FILE INF:
C   LABTB=LABEL OF DATA TABLE(40 CHARACTERS)
C   NQ=NUMBER OF QUADRATS(OBJECTS) IN LT
```

70

```
C    ITR=IF 0,DATA ENTERED BY SPECIES,ELSE BY OBJECTS
C         ALWAYS STORED BY SPECIES
C    IFPR=0,NO LISTING OF DATA PERFORMED
C    FROM FILE IND:
C    ROW=DATA STRING,ACCORDING TO SETS FORMAT IFM
C-------SUBROUTINE IASBC OR HETCALC:
C    IF NTB LT.0 AND NE.NQ,VECTOR NCOL(COLUMN NUMBERS OF THE
C    DATA(16I5)) HAS TO BE GIVEN(LAST LINE OR LINES IN INF)
C-------ONLY FOR DISCAN IN SUBROUTINE TAXLOOK:
C    FROM CARDS:
C    NTAXA=NUMBER OF REFERENCE TAXA
C    NOBTAX=TOTAL NUMBER OF OBJECTS IN REFERENCE TAXA
C    LOFMT=INPUT FORMAT FOR REFERENCE TAXA
C    FOR EACH OBJECT IN REFERENCE TAXON:
C    NTAXON=TAXON LABEL
C    ROW=OBJECT VECTOR(ACCORDING TO FORMAT LOFMT)
C    -------------------------------------------------------------
C    A  BLANK CARD TERMINATES SINFUN
C    *******************************************
      COMMON IDATA(20000),NCOL(1000),ROW(1000),RAW(1000),DISVEC(1000),
     1NTB,NDPW,NROWS,MOLT,IFPR,LOUTDIS,INF,IREAD
      DIMENSION MTITLE(3)
      PRINT 9999
9999  FORMAT(1HT)
      REWIND 1
      REWIND 2
      REWIND 3
      IBIANCO=10H
1     READ(5,2) (MTITLE(I),I=1,3),NTB,INF,IFUNCT,NDPW,NROWS,LOUTDIS
2     FORMAT(3A10,2I5,A10,6I5)
      IF(MTITLE(1)-IBIANCO) 3,1000,3
3     PRINT 4,(MTITLE(I),I=1,3),NTB,INF,IFUNCT,NDPW,NROWS,LOUTDIS
4     FORMAT(1H1,1X,3A10,* N.OF TABLES=*,I5/* INPUT FILE=*,I5/* REQUESTE
     1D I FUNCTION=*,A10/* N.DATA/WORD =*,I5/* N.OF ROWS=*,I5/* OUTFILE
     2FOR DISTANCE=*,I5)
      CALL STRUCOL
      IF(IREAD.GT.0) GO TO 1
      IF(IFUNCT.EQ.10HINF1     ) CALL IASBC(IFUNCT)
      IF(IFUNCT.EQ.10HINF1F    ) CALL IASBC(IFUNCT)
      IF(IFUNCT.EQ.10HINF2     ) CALL IASBC(IFUNCT)
      IF(IFUNCT.EQ.10HINF3     ) CALL IASBC(IFUNCT)
      IF(IFUNCT.EQ.10HINF3F    ) CALL IASBC(IFUNCT)
      IF(IFUNCT.EQ.10HDAB      ) CALL HETCALC(IFUNCT)
      IF(IFUNCT.EQ.10HSINGLEHET ) CALL HETCALC(IFUNCT)
      IF(IFUNCT.EQ.10HCLACOMP  ) CALL DOUBLEI(IFUNCT)
      IF(IFUNCT.EQ.10HDISCAN   ) CALL TAXLOOK(IFUNCT)
      GO TO 1
```

```
1000   PRINT 1001
1001   FORMAT(*0  SINFUN TERMINATED*)
       STOP
       END
       SUBROUTINE STRUCOL
       COMMON IDATA(20000),NCOL(1000),ROW(1000),RAW(1000),DISVEC(1000),
      1NTB,NDPW,NROWS,MOLT,IFPR,LOUTDIS,INF,IREAD
       DIMENSION LABTB(4),IFM(7),LFM(7)
       IREAD=0
         LAFUNZ=10HPACK
         LIMDIG=15
         NS=NROWS
       IF(NDPW.EQ.2) LIMDIG=10
       IF(NDPW.EQ.4) LIMDIG=5
       IF(NDPW.EQ.10) LIMDIG=2
       READ(5,1) (IFM(I),I=1,7),NDIG,NDEC
       READ(5,14)(LFM(I),I=1,7),IND
       IF(IND.LE.0) IND=INF
1      FORMAT(7A10,2I5)
14     FORMAT(7A10,I10)
       PRINT 9,(IFM(I),I=1,7),NDIG,NDEC
       PRINT19,(LFM(I),I=1,7),IND
9      FORMAT(*0*,7A10/* NDIG=*,I5,* NDEC=*,I5)
19     FORMAT(* LIST FORMAT *,7A10/* TABLES INPUT FILE *,I5)
       NINT=NDIG-NDEC
       IF(NDIG.LE.LIMDIG) GO TO 2
       IF(NINT.GT.LIMDIG) IREAD=1
       NDEC=LIMDIG-NINT
       NDIG=LIMDIG
2      MOLT=10**NDEC
       PRINT 3
3      FORMAT(*0    DATA LIST*)
       NTAB=NTB
         IF(NTB.LT.0) NTAB=1
       NPART=1
       DO 10 M=1,NTAB
       READ(INF,11) (LABTB(MU),MU=1,4),NQ,ITR,IFPR
11     FORMAT(4A10,8I5)
       PRINT 12,M,(LABTB(MU),MU=1,4),NQ,ITR,IFPR
12     FORMAT(*0  TAB.N.=*,I5,*  *,4A10/* NQ=*,I5,* ITR=*,I5,* IFPR=*,I5)
       NCOL(M)=NQ
       IF(ITR.GT.0) GO TO 4
       DO 5 I=1,NS
       READ(IND,IFM) (ROW(J),J=1,NQ)
       IF(IFPR.GT.0) PRINT LFM,(ROW(J),J=1,NQ)
       IF(IREAD.GT.0) GO TO 5
```

```
      II=I-1
        NPOS=II*NQ+NPART
        LPAS=1
      CALL PAUNPA(ROW,IDATA,LAFUNZ,IREAD,NQ,MOLT,LPAS,NPOS,NDPW)
5     CONTINUE
      GO TO 13
4     DO 6 I=1,NQ
      READ(IND,IFM) (ROW(J),J=1,NS)
      IF(IREAD.GT.0) GO TO 6
      IF(IFPR.GT.0) PRINT LFM,(ROW(J),J=1,NS)
      NPOS=I+NPART-1
        LPAS=NQ
        CALL PAUNPA(ROW,IDATA,LAFUNZ,IREAD,NS,MOLT,LPAS,NPOS,NDPW)
6     CONTINUE
13    NPART=NPART+NS*NQ
10    CONTINUE
      RETURN
      END
      SUBROUTINE PAUNPA(VEC,IDATA,LAFUNZ,IREAD,LUVEC,MOLT,LPAS,NPOS,NDW)
      DIMENSION VEC(1),IDATA(1),MASK(6)
      LPA=10HPACK
        LUNPA=10HUNPACK
      FMOLT=MOLT
      IF(NDW.EQ.10) GO TO 3
      IF(NDW.EQ.1) GO TO 5
      NPOS=NPOS-1
      MASK(1)=777770000000000000000B
        MASK(2)=7777700000000000B
      MASK(3)=7777700000B
        MASK(4)=77777B
      MASK(5)=7777777777770000000000B
        MASK(6)=7777777777B
      MAXDT=NDW*20000
        LIMSUP=32767
        IF(NDW.EQ.2)LIMSUP=1073741823
      IF(LAFUNZ.EQ.LUNPA) GO TO 2
      DO 11  I=1,LUVEC
      IDAT=VEC(I)*FMOLT+0.5
      IF(IDAT.LE.LIMSUP) GO TO 12
      GO TO 7
12    NW=NPOS/NDW+1
        IPOS=MOD(NPOS,NDW)
        LBIT=45-IPOS*15
      IF(NDW.EQ.2) LBIT=30-IPOS*30
```

```
          IDAT=SHIFT(IDAT,LBIT)
       LAMASK=IPOS+1
          IF(NDW.EQ.2) LAMASK=IPOS+5
       IDAT=AND(IDAT,MASK(LAMASK))
          IDATA(NW)=OR(IDATA(NW),IDAT)
       NPOS=NPOS+LPAS
          IF(NPOS.LE.MAXDT) GO TO 11
       PRINT 14,MAXDT
14     FORMAT(*  PAUNPA-MORE THAN  *,I7,*  DATA IN INPUT*)
       IREAD=1
       RETURN
11     CONTINUE
       RETURN
2      DO 21 I=1,LUVEC
       NW=NPOS/NDW+1
          IPOS=MOD(NPOS,NDW)
          LAMASK=IPOS+1
       IF(NDW.EQ.2) LAMASK=IPOS+5
          IDAT=AND(IDATA(NW),MASK(LAMASK))
       MBIT=-(45-IPOS*15)
          IF(NDW.EQ.2) MBIT=-(30-IPOS*30)
       IDAT=SHIFT(IDAT,MBIT)
          MASKUN=MASK(4)
       IF(NDW.EQ.2) MASKUN=MASK(6)
          IDAT=AND(IDAT,MASKUN)
       VEC(I)=FLOAT(IDAT)/FMOLT
          NPOS=NPOS+LPAS
21     CONTINUE
       RETURN
3      IF(LAFUNZ.EQ.LUNPA) GO TO 4
       DO 31 I=1,LUVEC
       LOVEC=VEC(I)*FMOLT+0.5
       IF(LOVEC.GT.63) GO TO 7
       CALL MOVCHR(10,LOVEC,NPOS,IDATA)
31     NPOS=NPOS+LPAS
       RETURN
4      DO 41 I=1,LUVEC
                        LOVEC=0
          CALL MOVCHR(NPOS,IDATA,10,LOVEC)
       VEC(I)=FLOAT(LOVEC)/FMOLT
41     NPOS=NPOS+LPAS
       RETURN
5      IF(LAFUNZ.EQ.LUNPA) GO TO 6
       DO 51 I=1,LUVEC
       LOVEC=VEC(I)*FMOLT+0.5
```

```
      IF(LOVEC.GT.(10**15)) GO TO 7
      IDATA(NPOS)=LOVEC
51    NPOS=NPOS+LPAS
      RETURN
6     DO 61 I=1,LUVEC
      VEC(I)=FLOAT(IDATA(NPOS))/FMOLT
61    NPOS=NPOS+LPAS
      RETURN
7     PRINT 13,IDAT,LIMSUP
13    FORMAT(*  PAUNPA-DATUM *,I12,*  EXCEEDS LIMIT *,I11)
      IREAD=1
      RETURN
      END
      SUBROUTINE IASBC(IFUNCT)
      COMMON IDATA(20000),NCOL(1000),ROW(1000),RAW(1000),DISVEC(1000),
     1NTB,NDPW,NROWS,MOLT,IFPR,LOUTDIS,INF,IREAD
      PRINT 6,IFUNCT
6     FORMAT(*0   *,A10,*  MATRIX*)
      IUMP=0
      IF(NTB.GT.0) GO TO 1
        IUMP=NCOL(1)
      NTB=-NTB
        IF(NTB.EQ.NCOL(1)) GO TO 7
      READ(5,2) (NCOL(I),I=1,NTB)
2     FORMAT(16I5)
      PRINT  10,(NCOL(I),I=1,NTB)
10    FORMAT(* NCOL  *,20I6)
      GO TO 1
7     IF(IFUNCT.EQ.10HINF3     .OR.IFUNCT.EQ.10HINF3F    ) GO TO 1
      DO 8 J=1,NTB
8     NCOL(J)=1
1     DO 3 I=2,NTB
      IM1=I-1
      DO 4 K=1,IM1
      IM=I
        IIM=K
      IF(IFUNCT.EQ.10HINF1     )DISVEC(K)=SINF1(ROW,RAW,NROWS,IM,IIM,
     1                               NDPW,NCOL,IDATA,MOLT,IUMP)
      IF(IFUNCT.EQ.10HINF1F    )DISVEC(K)=FINF1(ROW,RAW,NROWS,IM,IIM,
     1                               NDPW,NCOL,IDATA,MOLT,IUMP)
      IF(IFUNCT.EQ.10HINF2     )DISVEC(K)=SINF2(ROW,RAW,NROWS,IM,IIM,
     1                               NDPW,NCOL,IDATA,MOLT,IUMP)
      IF(IFUNCT.EQ.10HINF3     )DISVEC(K)=SINF3(ROW,RAW,NROWS,IM,IIM,
     1                               NDPW,NCOL,IDATA,MOLT)
      IF(IFUNCT.EQ.10HINF3F    )DISVEC(K)=FINF3(ROW,RAW,NROWS,IM,IIM,
     1                               NDPW,NCOL,IDATA,MOLT)
```

```
4       CONTINUE
        IF(LOUTDIS.NE.0) WRITE(LOUTDIS,5)(DISVEC(J),J=1,IM1)
        PRINT 9,I
9       FORMAT(* ROW  *,I5)
3       PRINT 5,(DISVEC(J),J=1,IM1)
5       FORMAT(5X,10G12.5)
        RETURN
        END
        FUNCTION SINF1(AMAT,BMAT,NRA,NA,NB,NDPW,NCOL,IDATA,MOLT,IUMP)
        DIMENSION AMAT(1),BMAT(1),NCOL(1),IDATA(1)
        LRA=NCOL(NA)
          LRB=NCOL(NB)
          NRB=NRA
          LCA=NRA
        LCB=NRB
          SINF1=0.
          LAF=10HUNPACK
          NPOSA=1
        NPOSB=1
          IR=0
          LP=1
          NAM1=NA-1
          NBM1=NB-1
        LARG=NRA
          IF(IUMP.GT.0) LARG=1
        FLRA=LRA
          FLRB=LRB
          FLRAB=FLRA+FLRB
          SINF1M=0.
        DO 4 J=1,NAM1
4       NPOSA=NPOSA+LARG*NCOL(J)
        IF(NB.EQ.1) GO TO 7
        DO 5 J=1,NBM1
5       NPOSB=NPOSB+LARG*NCOL(J)
7       DO 1 I=1,NRA
        TOTA=0.
          TOTB=0.
          TOTAB=0.
          IPOS=NPOSA
          IPO=NPOSA
        CALL PAUNPA(AMAT,IDATA,LAF,IR,LRA,MOLT,LP,IPOS,NDPW)
        NPOSA=IPO +LRA
          IF(IUMP.GT.0) NPOSA=IPO+IUMP
        VMAX=-1.E10
        DO 8 IC=1,LRA
        IF(AMAT(IC).GT.VMAX) VMAX=AMAT(IC)
```

```
8       CONTINUE
        DO 2 J=1,LRA
2       TOTA=TOTA+AMAT(J)
        IF(TOTA.EQ.0.) TOTA=1.E-9
        IPOS=NPOSB
          IPO=NPOSB
        CALL PAUNPA(BMAT,IDATA,LAF,IR,LRB,MOLT,LP,IPOS,NDPW)
        NPOSB=IPO+LRB
          IF(IUMP.GT.0) NPOSB=IPO+IUMP
        DO 9 IC=1,LRB
        IF(BMAT(IC).GT.VMAX) VMAX=BMAT(IC)
9       CONTINUE
        DO 3 J=1,LRB
3       TOTB=TOTB+BMAT(J)
        IF(TOTB.EQ.0.) TOTB=1.E-9
        TOTAB=TOTA+TOTB
        DIJK=TOTA*ALOG(TOTAB/TOTA)+TOTB*ALOG(TOTAB/TOTB)
        DIMJK=VMAX*(FLRA*ALOG(FLRAB/FLRA)+FLRB*ALOG(FLRAB/FLRB))
        SINF1=SINF1+DIJK
1       SINF1M=SINF1M+DIMJK
        SINF1=SINF1/SINF1M
        RETURN
        END
        FUNCTION FINF1(AMAT,BMAT,NRA,NA,NB,NDPW,NCOL,IDATA,MOLT,IUMP)
        DIMENSION AMAT(1),BMAT(1),NCOL(1),IDATA(1)
        LRA=NCOL(NA)
          LRB=NCOL(NB)
          NRB=NRA
          LCA=NRA
        LCB=NRB
          FINF1=0.
          LAF=10HUNPACK
          NPOSA=1
        NPOSB=1
          IR=0
          LP=1
          NAM1=NA-1
          NBM1=NB-1
        LARG=NRA
          IF(IUMP.GT.0) LARG=1
          LRAB=LRA+LRB
        FLRAB=LRAB
          FLRA=LRA
          FLRB=LRB
          TINF=0.
        COST=FLRA*ALOG(FLRAB/FLRA)+FLRB*ALOG(FLRAB/FLRB)
```

```
      DO 40 J=1,NAM1
40    NPOSA=NPOSA+LARG*NCOL(J)
      IF(NB.EQ.1) GO TO 70
      DO 50 J=1,NBM1
50    NPOSB=NPOSB+LARG*NCOL(J)
70    DO 1 I=1,NRA
      IPOS=NPOSA
        IPO=NPOSA
      CALL PAUNPA(AMAT,IDATA,LAF,IR,LRA,MOLT,LP,IPOS,NDPW)
      NPOSA=IPO+LRA
        IF(IUMP.GT.0) NPOSA=IPO+IUMP
      IPOS=NPOSB
        IPO=NPOSB
      CALL PAUNPA(BMAT,IDATA,LAF,IR,LRB,MOLT,LP,IPOS,NDPW)
      NPOSB=IPO+LRB
        IF(IUMP.GT.0) NPOSB=IPO+IUMP
      RINF=0.
      DO 60  JJ=1,LRA
      IF(AMAT(JJ).LE.0.) AMAT(JJ)=1.E-9
60    CONTINUE
      DO 65  JJ=1,LRB
      IF(BMAT(JJ).LE.0.) BMAT(JJ)=1.E-9
65    CONTINUE
      DO 2 J=1,LRA
      IFA=IFB=0
      AR=AMAT(J)
      IF(AR.LE.0.) GO TO 2
      IFB=NUCAR(BMAT,AR,LRB)
      IF(IFB.EQ.0) GO TO 2
      IFA=NUCAR(AMAT,AR,LRA)
      FA=IFA
        FB=IFB
        FAB=FA+FB
      CINF=FA*ALOG(FA/FAB)+FB*ALOG(FB/FAB)
      RINF=RINF+CINF
2     CONTINUE
1     TINF=TINF+RINF+COST
      FINF1=ABS(TINF)
      RETURN
      END
      FUNCTION NUCAR(VEC,CAR,LVEC)
      DIMENSION VEC(1)
      NUCAR=0
      DO 1 J=1,LVEC
      IF(VEC(J).LE.0.) GO TO 1
      IF(VEC(J).NE.CAR) GO TO 1
```

```
      VEC(J)=-VEC(J)
      NUCAR=NUCAR+1
1     CONTINUE
      RETURN
      END
      FUNCTION SINF2(AMAT,BMAT,NRA,NA,NB,NDPW,NCOL,IDATA,MOLT,IUMP)
      DIMENSION AMAT(1),BMAT(1),NCOL(1),IDATA(1)
      LRA=NCOL(NA)
        LRB=NCOL(NB)
        NRB=NRA
        LCA=NRA
      LCB=NRB
        LAF=10HUNPACK
        IR=0
        LP=1
      SINF2=RA=RB=RAB=TA=TB=TAB=0.
      NPOSA=1
        NPOSB=1
        NAM1=NA-1
        NBM1=NB-1
      LARG=NRA
        IF(IUMP.GT.0) LARG=1
      DO 8 J=1,NAM1
8     NPOSA=NPOSA+LARG*NCOL(J)
      IF(NB.EQ.1) GO TO 13
      DO 9 J=1,NBM1
9     NPOSB=NPOSB+LARG*NCOL(J)
13    DO 1 I=1,NRA
      IPOS=NPOSA
        IPO=NPOSA
      CALL PAUNPA(AMAT,IDATA,LAF,IR,LRA,MOLT,LP,IPOS,NDPW)
      NPOSA=IPO+LRA
        IF(IUMP.GT.0) NPOSA=IPO+IUMP
      TOTA=TOTB=TOTAB=0.
      DO 2 J=1,LRA
      AEL=AMAT(J)
        IF(AEL.EQ.0.)AEL=1.E-9
        TOTA=TOTA+AEL
2     TA=TA+AEL
      RA=RA-TOTA*ALOG(TOTA)
        IPOS=NPOSB
       IPO=NPOSB
      CALL PAUNPA(BMAT,IDATA,LAF,IR,LRB,MOLT,LP,IPOS,NDPW)
      NPOSB=IPO+LRB
        IF(IUMP.GT.0) NPOSB=IPO+IUMP
      DO 3 J=1,LRB
```

```
      BEL=BMAT(J)
        IF(BEL.EQ.0.)BEL=1.E-9
        TOTB=TOTB+BEL
3     TB=TB+BEL
      RB=RB-TOTB*ALOG(TOTB)
        TOTAB=TOTA+TOTB
        TAB=TAB+TOTAB
1     RAB=RAB-TOTAB*ALOG(TOTAB)
      ABMU=RAB+TAB*ALOG(TAB)
      AMU=RA+TA*ALOG(TA)
      BMU=RB+TB*ALOG(TB)
      SINF2=ABS(ABMU-AMU-BMU)
      RETURN
      END
      FUNCTION SINF3(AMAT,BMAT,NRA,NA,NB,NDPW,NCOL,IDATA,MOLT)
      DIMENSION AMAT(1),BMAT(1),IDATA(1),NCOL(1)
      NRB=NRA
        LP=NCOL(1)
        IR=0
        SINF3=0.
        LAF=10HUNPACK
      TRIG=TCOL1=TCOL2=SJK=SRIG=SCOL=0.
      CALL PAUNPA(AMAT,IDATA,LAF,IR,NRA,MOLT,LP,NA,NDPW)
      CALL PAUNPA(BMAT,IDATA,LAF,IR,NRB,MOLT,LP,NB,NDPW)
      DO 1 I=1,NRA
      IF(AMAT(I).EQ.0.)AMAT(I)=1.E-9
       IF(BMAT(I).EQ.0.)BMAT(I)=1.E-9
      TRIG=AMAT(I)+BMAT(I)
      TCOL1=TCOL1+AMAT(I)
        TCOL2=TCOL2+BMAT(I)
      SJK=-(AMAT(I)*ALOG(AMAT(I))+BMAT(I)*ALOG(BMAT(I)))+SJK
1     SRIG=-TRIG*ALOG(TRIG)+SRIG
      SCOL=-TCOL1*ALOG(TCOL1)-TCOL2*ALOG(TCOL2)
      SINF3=ABS(2.*SJK-SRIG-SCOL)
      RETURN
      END
      FUNCTION FINF3(AMAT,BMAT,NRA,NA,NB,NDPW,NCOL,IDATA,MOLT)
      DIMENSION AMAT(1),BMAT(1),IDATA(1),NCOL(1)
      NRB=NRA
        LP=NCOL(1)
        IR=0
        FINF3=0.
        LAF=10HUNPACK
      TRIG=TCOL=TJOINT=0.
      CALL PAUNPA(AMAT,IDATA,LAF,IR,NRA,MOLT,LP,NA,NDPW)
      CALL PAUNPA(BMAT,IDATA,LAF,IR,NRB,MOLT,LP,NB,NDPW)
```

```
      DO 1 IC=1,NRA
      IF(AMAT(IC).LE.0.) AMAT(IC)=1.E-9
      IF(BMAT(IC).LE.0.) BMAT(IC)=1.E-9

1     CONTINUE
      DO 2 IC=1,NRA
      CAR=AMAT(IC)
        IF(CAR.LE.0.) GO TO 2
      FNC=NUCAR(AMAT,CAR,NRA)
        TRIG=TRIG+FNC*ALOG(FNC)
2     CONTINUE
      DO 3 IC=1,NRB
      CAR=BMAT(IC)
        IF(CAR.LE.0.) GO TO 3
      FNC=NUCAR(BMAT,CAR,NRB)
        TCOL=TCOL+FNC*ALOG(FNC)
3     CONTINUE
      DO 4 IC=1,NRA
      AMAT(IC)=ABS(AMAT(IC))
4     BMAT(IC)=ABS(BMAT(IC))
      DO 5 IC=1,NRA
      PR=AMAT(IC)
        SEC=BMAT(IC)
        IF(PR.LE.0.) GO TO 5
      FNUP=NUPAIRS(AMAT,BMAT,PR,SEC,NRA)
      TJOINT=TJOINT-2.*FNUP*ALOG(FNUP)
5     CONTINUE
      FINF3=TJOINT+TRIG+TCOL
      RETURN
      END
      FUNCTION NUPAIRS(A,B,PR,SEC,L)
      DIMENSION A(1),B(1)
      NUPAIRS=0
      DO 1 IC=1,L
      AA=A(IC)
        BB=B(IC)
        IF(AA.NE.PR.OR.BB.NE.SEC) GO TO 1
      NUPAIRS=NUPAIRS+1
        A(IC)=-A(IC)
        B(IC)=-B(IC)
1     CONTINUE
      RETURN
      END
      SUBROUTINE HETCALC(IFUNCT)
      COMMON IDATA(20000),NCOL(1000),ROW(1000),RAW(1000),DISVEC(1000),
     1NTB,NDPW,NROWS,MOLT,IFPR,LOUTDIS,INF,IREAD
```

```
      IUMP=0
      IF(IFUNCT.EQ.10HSINGLEHET ) GO TO 7
      PRINT 6,IFUNCT
6     FORMAT(*0   *,A10,*  MATRIX*)
      IF(NTB.GT.0) GO TO 1
        IUMP=NCOL(1)
      NTB=-NTB
        IF(NTB.EQ.NCOL(1)) GO TO 9
      READ(5,2) (NCOL(I),I=1,NTB)
2     FORMAT(16I5)
      PRINT 12,(NCOL(I),I=1,NTB)
12    FORMAT(* NCOL  *,20I6)
      GO TO 1
9     DO 10 J=1,NTB
10    NCOL(J)=1
1     DO 3 I=2,NTB
      IM1=I-1
      DO 4 K=1,IM1
      MA=I
        MB=K
4     DISVEC(K)=DAB(ROW,RAW,NROWS,MA,MB,NDPW,NCOL,IDATA,MOLT,IUMP)
      IF(LOUTDIS.NE.0) WRITE(LOUTDIS,5) (DISVEC(J),J=1,IM1)
      PRINT 11,I
11    FORMAT(* ROW  *,I5)
3     PRINT 5,(DISVEC(J),J=1,IM1)
5     FORMAT(5X,10G12.5)
      RETURN
7     HET=HETERO(IDATA,NCOL(1),NROWS,ROW,MOLT,NDPW)
      PRINT 8,HET
8     FORMAT(*0    TABLE HETEROGENEITY=*,G12.5)
      RETURN
      END
      FUNCTION DAB(AMAT,BMAT,NRA,NA,NB,NDPW,NCOL,IDATA,MOLT,IUMP)
      DIMENSION AMAT(1),BMAT(1),NCOL(1),IDATA(1)
      LCA=NRA
        NCA=NCOL(NA)
        NCB=NCOL(NB)
        NRB=NRA
      LCB=LCA
        LAF=10HUNPACK
        DAB=0.
        IR=0
        NPOSA=1
      NPOSB=1
        NAM1=NA-1
        NBM1=NB-1
```

```
      LARG=NRA
        IF(IUMP.GT.0) LARG=1
      DO 4 J=1,NAM1
4     NPOSA=NPOSA+LARG*NCOL(J)
      IF(NB.EQ.1) GO TO 7
      DO 5 J=1,NBM1
5     NPOSB=NPOSB+LARG*NCOL(J)
7     DO 1 I=1,NRA
      TOTA=0.
        TOTB=0.
        LP=1
        IPOS=NPOSA
        IPO=NPOSA
      CALL PAUNPA(AMAT,IDATA,LAF,IR,NCA,MOLT,LP,IPOS,NDPW)
      NPOSA=IPO+NCA
        IF(IUMP.GT.0) NPOSA=IPO+IUMP
      DO 2 J=1,NCA
2     TOTA=TOTA+AMAT(J)
      IF(TOTA.EQ.0.) TOTA=1.E-9
        XAM=TOTA/FLOAT(NCA)
        IPOS=NPOSB
      IPO=NPOSB
      CALL PAUNPA(BMAT,IDATA,LAF,IR,NCB,MOLT,LP,IPOS,NDPW)
      NPOSB=IPO+NCB
        IF(IUMP.GT.0) NPOSB=IPO+IUMP
      DO 3 J=1,NCB
3     TOTB=TOTB+BMAT(J)
      IF(TOTB.EQ.0.) TOTB=1.E-9
        XBM=TOTB/FLOAT(NCB)
      XABM=(TOTA+TOTB)/FLOAT(NCA+NCB)
      DABI=TOTA*ALOG(XAM/XABM)+TOTB*ALOG(XBM/XABM)
1     DAB=DAB+DABI
      DAB=ABS(DAB)
      RETURN
      END
      FUNCTION HETERO(IDATA,LR,LC,XBUF,MOLT,NDPW)
      DIMENSION IDATA(1),XBUF(1)
      NR=LC
        NC=LR
        LAF=10HUNPACK
        IR=0
        LP=1
      HETERO=0.
      DO 1 I=1,NR
      IPOS=(I-1)*LR+1
      CALL PAUNPA(XBUF,IDATA,LAF,IR,LR,MOLT,LP,IPOS,NDPW)
```

```
      DEPINFI=SINFOX(XBUF,NC,1)-SINFOX(XBUF,NC,0)
1     HETERO=HETERO+DEPINFI
      HETERO=2.*HETERO
      RETURN
      END
      FUNCTION SINFOX(X,NX,ISW)
      DIMENSION X(1)
      SINFOX=0.
        TOTX=0.
      DO 1 I=1,NX
1     TOTX=TOTX+X(I)
      IF(TOTX.EQ.0.) TOTX=1.E-9
        XMED=TOTX/FLOAT(NX)
      DO 2 I=1,NX
      XI=X(I)
        IF(XI.EQ.0.)XI=1.E-9
        PIJ=XI/TOTX
      IF(ISW.EQ.1)PIJ=XMED/TOTX
        DS=-X(I)*ALOG(PIJ)
2     SINFOX=SINFOX+DS
      RETURN
      END
      SUBROUTINE DOUBLEI(IFUNCT)
      COMMON IDATA(20000),NCOL(1000),ROW(1000),RAW(1000),DISVEC(1000),
     1NTB,NDPW,NROWS,MOLT,IFPR,LOUTDIS,INF,IREAD
      DIMENSION DECOMP(1000)
      NC=NCOL(1)
        LPR=1
        LPC=NC
                  LAF=10HUNPACK
      IR=0
        GENTOT=0.
        TROWI=0.
        GIOINT=0.
      DO 1 J=1,NROWS
      TOTR=0.
        IPOSR=1+(J-1)*NC
      CALL PAUNPA(ROW,IDATA,LAF,IR,NC,MOLT,LPR,IPOSR,NDPW)
      DO 2 I=1,NC
      ROWI=ROW(I)
        IF(ROWI.EQ.0.)ROWI=1.E-9
      GIOINT=GIOINT+ROWI*ALOG(ROWI)
2     TOTR=TOTR+ROWI
      DECOMP(J)=TOTR
      TROWI=TROWI-TOTR*ALOG(TOTR)
1     GENTOT=GENTOT+TOTR
```

```
      TABI=GENTOT*ALOG(GENTOT)
        TCOLI=0.
      BETR=2.*(TABI+TROWI)
       WIR=-2.*(TROWI+GIOINT)
      DO 3 J=1,NC
      TOTC=0.
        IPOSC=J
      CALL PAUNPA(RAW,IDATA,LAF,IR,NROWS,MOLT,LPC,IPOSC,NDPW)
      DO 4 I=1,NROWS
      COLI=RAW(I)
        IF(COLI.EQ.0.) COLI=1.E-9
4     TOTC=TOTC+COLI
      DISVEC(J)=TOTC
3     TCOLI=TCOLI-TOTC*ALOG(TOTC)
      BETC=2.*(TABI+TCOLI)
       WIC=-2.*(TCOLI+GIOINT)
      DUEI=2.*(GIOINT+TABI+TROWI+TCOLI)
      INC=(NROWS-1)*(NC-1)
        ALLITO=DUEI
      CALL REPRO(BO,INC,ALLITO)
        BOM=100.-BO
      PRINT 5,DUEI,INC,BOM
5     FORMAT(*0  2I VALUE   *,G12.5,* FOR *,I5,* DEGREES OF FREEDOM *,*
     1WITH CHISQUARE PROBABILITY *,F6.1)
      PRINT66,BETR,WIR,BETC,WIC
66    FORMAT(14H0BETWEEN ROWS ,G12.5/13H WITHIN ROWS ,G12.5/
     117H0BETWEEN COLUMNS ,G12.5/16H WITHIN COLUMNS ,G12.5)
      INC=NC-1
      PRINT 12,INC
12    FORMAT(6H0  ROW,12H   2I(ROW)  ,10H  DEG.FR.=,I5,8H  P(X2)            )
      DO 6 J=1,NROWS
      IPOSR=1+(J-1)*NC
       AIRIG=0.
       AIRC=0.
      CALL PAUNPA(ROW,IDATA,LAF,IR,NC,MOLT,LPR,IPOSR,NDPW)
      DO 7 I=1,NC
      ROWI=ROW(I)
       IF(ROWI.LE.0.)ROWI=1.E-9
      AIRIG=AIRIG+ROWI*ALOG(ROWI)
7     AIRC=AIRC+ROWI*ALOG(DISVEC(I))
6     RAW(J)=2.*(AIRIG+DECOMP(J)*ALOG(GENTOT)-AIRC-DECOMP(J)*ALOG(DECOMP
     1(J)))
      NUMGIR=1
       JMAX=1
11    AMAX=-1.E20
```

```
      DO 8 J=1,NROWS
      IF(RAW(J).LE.AMAX)GO TO 8
       AMAX=RAW(J)
       JMAX=J
8     CONTINUE
      ALLITO=RAW(JMAX)
       CALL REPRO(BO,INC,ALLITO)
       BOM=100.-BO
      PRINT 9,JMAX,RAW(JMAX),BOM
9     FORMAT(1X,I5,G12.5,15X,F6.1)
      RAW(JMAX)=-JMAX
       NUMGIR=NUMGIR+1
       IF(NUMGIR.GT.NROWS)GO TO 10
      GO TO 11
10    INC=NROWS-1
       PRINT 13,INC
13    FORMAT(6H0 COL.,12H  2I(COL.)  ,10H DEG.FR.=,I5,8H  P(X2)         )
      DO 14 J=1,NC
      IPOSC=J
       AIRIG=0.
       AIRC=0.
      CALL PAUNPA(ROW,IDATA,LAF,IR,NROWS,MOLT,LPC,IPOSC,NDPW)
      DO 15 I=1,NROWS
      COLI=ROW(I)
       IF(COLI.LE.0.)COLI=1.E-9
      AIRIG=AIRIG+COLI*ALOG(COLI)
15    AIRC=AIRC+COLI*ALOG(DECOMP(I))
14    RAW(J)=2.*(AIRIG+DISVEC(J)*ALOG(GENTOT)-AIRC-DISVEC(J)*ALOG(DISVEC
     1(J)))
      NUMGIR=1
      JMAX=1
18    AMAX=-1.E20
      DO 16 J=1,NC
      IF(RAW(J).LE.AMAX)GO TO 16
       AMAX=RAW(J)
       JMAX=J
16    CONTINUE
      ALLITO=RAW(JMAX)
       CALL REPRO(BO,INC,ALLITO)
       BOM=100.-BO
      PRINT 9,JMAX,RAW(JMAX),BOM
       RAW(JMAX)=-JMAX
       NUMGIR=NUMGIR+1
      IF(NUMGIR.LE.NC) GO TO 18
      RETURN
      END
```

86

```
      SUBROUTINE REPRO(BO,INC,ALLITO)
C---IT PRODUCES NULL-HYPOTHESIS PROBABILITY FOR A GIVEN CHI-SQUARE.
      DIMENSION FI(1001)
      N=INC
        COVAL=ALLITO
        NPAS=1000
      IF(N.EQ.2) GO TO 1000
      PIGR=3.141592654
      A=0.
        FN=FLOAT(N)
        RADN=SQRT(4.*FN)
        AB=FN-7.*RADN
      IF(AB.GT.A) A=AB
        IF(N.EQ.1) A=0.0642
        QPAS=FLOAT(NPAS)
      B=FN+7.*RADN
       H=(B-A)/QPAS
       FI(1)=0.
       IF(N.EQ.1)FI(1)=0.1999
      X=A
        GAML=GAM(N)
        TI=F(X,FN,GAML)
      IF(COVAL.LT.A.AND.N.NE.1) GO TO 7
        IF(COVAL.GT.B)GO TO 8
      DO 1 I=1,NPAS
      X=X+H
      TII=F(X,FN,GAML)
      DIF=(TII+TI)/2.*H
      TI=TII
1     FI(I+1)=FI(I)+DIF
      IF(N.GT.2) GO TO 2
      IF(COVAL.GT.0.0642) GO TO 2
      BO=(1.-SQRT(2.*COVAL/PIGR))*100.
      GO TO 3
2     DO 4 I=1,NPAS
      XI=A+H*FLOAT(I-1)
        XII=XI+H
      IF(COVAL.GE.XI.AND.COVAL.LE.XII) GO TO 5
4     CONTINUE
      GO TO 3
7     BO=100.
        GO TO 3
8     BO=0.
        GO TO 3
5     DFI=FI(I+1)-FI(I)
        BO=(1.-(DFI*(COVAL-XI)/H+FI(I)))*100.
```

```
3      CONTINUE
       RETURN
1000   BO=(EXP(-COVAL/2.))*100.
       RETURN
       END
       FUNCTION F(X,FN,GAML)
       F=0.
       IF(X.LE.0.) RETURN
       ALOGF=-0.6931471806+(FN/2.-1.)*ALOG(X/2.)-X/2.-GAML
       F=EXP(ALOGF)
       RETURN
       END
       FUNCTION GAM(N)
       RAPIGR=1.144729886/2.
         ILIM=N/2
       IF(MOD(N,2).EQ.0) GO TO 1
       GAM=RAPIGR
       IF(N.EQ.1) RETURN
       GAMLN=0.
       DO 2 I=1,ILIM
2      GAMLN=GAMLN+ALOG(FLOAT(I)-0.5)
       GAM=GAMLN+RAPIGR
       RETURN
1      GAMLN=0.
       ILIM=ILIM-1
       DO 3 I=1,ILIM
3      GAMLN=GAMLN+ALOG(FLOAT(I))
       GAM=GAMLN
       RETURN
       END
       SUBROUTINE TAXLOOK(IFUNCT)
       COMMON IDATA(20000),NCOL(1000),ROW(1000),RAW(1000),DISVEC(1000),
      1NTB,NDPW,NROWS,MOLT,IFPR,LOUTDIS,INF,IREAD
       DIMENSION LOFMT(7)
       INTEGER TAPET,TAPEO
       TAPET=2
         TAPEO=3
         REWIND TAPET
         REWIND  TAPEO
       READ(5,1) NTAXA,NOBTAX,(LOFMT(I),I=1,7)
1      FORMAT(2I5,7A10)
       PRINT 17,NTAXA,NOBTAX,(LOFMT(I),I=1,7)
17     FORMAT(20X,*---TAXLOOK---*/* NTAXA=*,I5/* NOBTAX=*,I5/1X,7A10)
       DO 2 I=1,NOBTAX
       READ( 5 ,LOFMT) NTAXON,(ROW(J),J=1,NROWS)
       WRITE(TAPET)       NTAXON,(ROW(J),J=1,NROWS)
```

88

```
      PRINT 15,I,NTAXON
15    FORMAT(* OBJECT N. *,I5,* REF.TAXON N. *,I5)
2     PRINT 16,(ROW(J),J=1,NROWS)
16    FORMAT(10X,10G12.5)
      NOB=NCOL(1)
      DO 3 I=1,NOB
      IPOS=I
        LP=NOB
        LAF=10HUNPACK
        IR=0
        VMIN=1.E20
      REWIND TAPET
      CALL PAUNPA(ROW,IDATA,LAF,IR,NROWS,MOLT,LP,IPOS,NDPW)
      DO 4 K=1,NOBTAX
      READ(TAPET)        NTAXON,(RAW(J),J=1,NROWS)
      DISI=0.
      DO 5 J=1,NROWS
      ROWJ=ROW(J)
        RAWJ=RAW(J)
        IF(ROWJ.EQ.0.)ROWJ=1.E-9
      IF(RAWJ.EQ.0.) RAWJ=1.E-9
      DI=ROWJ*ALOG(2.*ROWJ/(ROWJ+RAWJ))+RAWJ*ALOG(2.*RAWJ/(ROWJ+RAWJ))
5     DISI=DISI+DI
      IF(DISI.GE.VMIN) GO TO 4
      VMIN=DISI
        MINTAX=NTAXON
        NEIGHB=K
4       CONTINUE
      IOBCT=I
      INC=NROWS-1
      ALLITO=VMIN
      CALL REPRO(BO,INC,ALLITO)
      BOM=100.-BO
3     WRITE(TAPEO)MINTAX,NEIGHB,(ROW(J),J=1,NROWS),IOBCT
     1,VMIN,BOM
      REWIND TAPET
        WITHIN=0.
        SAMPINF=0.
      DO 6 I=1,NTAXA
      REWIND TAPEO
        TAXINF=0.
        NTAXI=0
      PRINT 18,I
18    FORMAT(*0 DATA-TAXON N. *,I5)
      DO 7 M=1,NROWS
      ROW(M)=0.
```

```
      RAW(M)=0,
7     DISVEC(M)=0,
      IFOUND=0
      DO 8 J=1,NOB
      READ(TAPEO) LOBTAX,ILVICIN,(DISVEC(M),M=1,NROWS),IOBCT
     1,CHIC,BOM
      IF(LOBTAX.NE.I) GO TO 8
      IFOUND=1
      PRINT 10,IOBCT,(DISVEC(M),M=1,NROWS)
10    FORMAT(*---DATA OBJECT N.*,I5/(1X,10G11.4))
      PRINT 11,ILVICIN,CHIC,BOM
11    FORMAT(* NEAREST OBJECT IN REFERENCE TAXA *,I5,
     1" X2=",G12.5," DIVERGENCE PROB(%)=",G11.4)
      NTAXI=NTAXI+1
      DO 9 M=1,NROWS
      IF(DISVEC(M).EQ.0.) DISVEC(M)=1.E-9
      ROW(M)=ROW(M)+DISVEC(M)*ALOG(DISVEC(M))
9     RAW(M)=RAW(M)+DISVEC(M)
8     CONTINUE
      IF(IFOUND.EQ.0) GO TO 6
      DO 12 M=1,NROWS
12    TAXINF=TAXINF+2.*(ROW(M)-RAW(M)*ALOG(RAW(M)/FLOAT(NTAXI)))
      LABTAX=I
      INC=(NTAXI-1)*(NROWS-1)
      ALLITO=TAXINF
      CALL REPRO(BO,INC,ALLITO)
      BIM=100.-BO
      PRINT 13,LABTAX,TAXINF,BIM
13    FORMAT(*0 TAXON *,I5,* HETEROGENEITY *,G12.5/
     1" WITH PROBABILITY(%)=",G11.4)
      WITHIN=WITHIN+TAXINF
6     CONTINUE
      SAMPINF=   HETERO(IDATA,NOB,NROWS,ROW,MOLT,NDPW)
      INC=(NROWS-1)*(NOB-1)
      ALLITO=SAMPINF
      CALL REPRO(BO,INC,ALLITO)
      BOM=100.-BO
      BETWEEN=(SAMPINF-WITHIN)/SAMPINF
      PRINT 14,SAMPINF,WITHIN,BETWEEN,BOM
14    FORMAT(*0 TOTAL SAMPLE HETEROGENEITY=*,G12.5/* TOTAL WITHIN-TAXA
     1HETEROGENEITY=*,G12.5/* RATIO (BETWEEN TAXA)/(TOTAL SAMPLE)=*,
     2G12.5/" TOTAL SAMPLE X2-PROB(%)=",G11.4)
      RETURN
      END
CNESTOF
      PROGRAM NESTOFL(INPUT=100,OUTPUT=100,TAPE5=INPUT,TAPE1=514,
```

90

```
      1TAPE2=514,TAPE3=514,TAPE4=514)
C----IN MAIN PROGRAM
C       ITIT=40CHARACTERS FOR TITLE,IF FIRST TEN BLANKS,RUN STOPS
C       NESTYP1=IF EQ. "HIERARCHIC",NESTED ANALISYS PERFORMED
C       NESTYP2=IF EQ. "BINARYTWOI",INTERACTION ANALISYS PERFORMED
C       NSPEC=NUMBER OF SPECIES(UP TO 1000)
C       NQUAD=NUMBER OF QUADRATS(UP TO 1000)
C       ITRASP=IF GT.0,DATA ENTERED BY RELEVES,ELSE BY SPECIES
C       (DATA ALWAYS STORED BY SPECIES)
C       IFPR=IF GT 0,DATA PRINTED
C       INF=INPUT FILE FOR DATA(5 FOR CARDS,1 OR 2 FOR TAPE)
C----    IN SUBROUTINE READATA
C       IFMT=INPUT FORMAT FOR DATA (8A10)
C       LFMT=FORMAT FOR PRINTING DATA (8A10)
C FROM FILE "INF":
C       DATAR=DATA TABLE(FORMAT IFMT)
C----    IN SUBROUTINE HIENEST AND/OR SUBROUTINE DUINEST
C       NLEV=NUMBER OF HIERARCHICAL LEVELS
C       NGR=NUMBER OF BLOCKS
C       NESDATA=POSITION OF FIRST OBJECT IN SECOND AND SUBSEQUENT BLOCKS
C  .  .  .  .  .  .  .  .  .  .  .  .  .  .  .  .  .  .  .  .  .  .  .
C    A BLANK CARD TERMINATES NESTOFL
C  .  .  .  .  .  .  .  .  .  .  .  .  .  .  .  .  .  .  .  .  .  .  .
C
      COMMON NESDATA(2000),DATAR(1000),SPECR(1000),INF
      DIMENSION ITIT(4)
      PRINT 9999
9999  FORMAT(1HT)
      REWIND1
        REWIND 2
        IBLANK=10H
      REWIND 3
        REWIND 4
5     READ(5,1) ITIT,NESTYP1,NESTYP2
1     FORMAT(6A10)
      IF(ITIT(1).EQ.IBLANK) GO TO 9998
      PRINT 2,ITIT,NESTYP1,NESTYP2
2     FORMAT(1H1,3X,4A10,3X,A10,3X,A10)
      READ(5,3) NSPEC,NQUAD,ITRASP,IFPR,INF
3     FORMAT(16I5)
      PRINT 4,NSPEC,NQUAD,ITRASP,IFPR,INF
4     FORMAT(* NSPEC NQUAD ITRASP IFPR INF*/1X,5I6)
      CALL READATA(ITRASP,IFPR,NSPEC,NQUAD,LBUF)
      IF(NESTYP1.EQ.10HHIERARCHIC) CALL HIENEST(NSPEC,NQUAD,LBUF)
      IF(NESTYP2.EQ.10HBINARYTWOI) CALL DUINEST(NSPEC,NQUAD,LBUF)
      GO TO 5
```

```
9998  PRINT 9997
9997  FORMAT(*0NESTOFL TERMINATED*)
      STOP
      END
      SUBROUTINE READATA(ITR,IPR,NSPEC,NQUAD,LBUF)
      COMMON NESDATA(2000),DATAR(1000),SPECR(1000),INF
      DIMENSION IFMT(8),LFMT(8)
      IND=INF
        IF(INF.EQ.5) IND=1
        LBUF=MOD(IND,2)+1
      READ(5,1) IFMT,LFMT
1     FORMAT(8A10)
      N=0
        IF(ITR.GT.0) GO TO 2
      DO 11 J=1,NSPEC
      READ(INF,IFMT) (DATAR(I),I=1,NQUAD)
      WRITE(LBUF) (DATAR(I),I=1,NQUAD)
      IF(IPR.LE.0) GO TO 11
        N=N+1
        PRINT 10,N
10    FORMAT(* DATA ROW N.=*,I5)
      PRINT LFMT,(DATAR(I),I=1,NQUAD)
11    CONTINUE
      RETURN
2     DO 20 J=1,NQUAD
      N=N+1
      READ(INF,IFMT) (DATAR(I),I=1,NSPEC)
      WRITE(LBUF) (DATAR(I),I=1,NSPEC)
      IF(IPR.LE.0) GO TO 20
        PRINT 10,N
      PRINT LFMT,(DATAR(I),I=1,NSPEC)
20    CONTINUE
      REWIND IND
      DO 21 J=1,NSPEC
      REWIND LBUF
      DO 22 JJ=1,NQUAD
      READ(LBUF) (SPECR(I),I=1,NSPEC)
22    DATAR(JJ)=SPECR(J)
21    WRITE(IND) (DATAR(I),I=1,NQUAD)
      LBUF=IND
      RETURN
      END
      SUBROUTINE HIENEST(NS,NQ,LB)
      COMMON NESDATA(2000),DATAR(1000),SHAME(1000),INF
      INTEGER DEFREDO
      FNQ=FLOAT(NQ)
```

```
       REWIND 3
      READ(5,1) NLEV
 1     FORMAT(16I5)
                                 IP=0
       NLEV1=NLEV-1
       PRINT3,NLEV
 3     FORMAT(*0NUMBER OF LEVELS=*,I5,* HIERARCHIC NESTING*)
       IF(NLEV1.LE.0) GO TO 1000
       DO 2 I=1,NLEV1
       IUHU=I+1
       READ(5,1) NGR
        PRINT4,NGR,IUHU
 4     FORMAT(1X,I5,* GROUPS IN LEVEL N.=*,I5)
       IP=IP+1
        NGR1=NGR-1
        NESDATA(IP)=-NGR1
       IP1=IP+1
        IP=IP+NGR1
        IF(IP.GT.2000) GO TO 2
       READ(5,1) (NESDATA(K),K=IP1,IP)
       PRINT5,(NESDATA(K),K=IP1,IP)
 5     FORMAT(* FIRST OBJECT IN GROUPS,BESIDES N.  1*,20I5)
 2     CONTINUE
1000   REWIND LB
       DO 10 I=1,NS
       PRINT 200,I
200    FORMAT(*0 SPECIES N.=*,I5)
       READ(LB) (DATAR(J),J=1,NQ)
        TG=0.
        IL=1
        PINF=0.
       DO 11 J=1,NQ
       DATA=DATAR(J)
        IF(DATA.LE.0.)DATA=1.E-10
       PINF=PINF+DATA*ALOG(DATA)
        SHAME(J)=DATA
11     TG=TG+DATA
       TGMED=TG/FNQ
        PINF=PINF-TG*ALOG(TGMED)
       NUGRU=NQ-1
        IF(NLEV1.LE.0)GO TO 10
       DO 12 K=1,NLEV1
       IL=K
        KI=K
        CALL OUS(SHAME,PANF,KI,NESDATA,NQ,NOGRU,TGMED)
       CREPA=PINF-PANF
```

```
       CREPA=ABS(CREPA)
         DEFREDO=NUGRU-NOGRU
       PINF=ABS(PANF)
       NUGRU=NOGRU
         CALL REPRO(BO,DEFREDO,CREPA)
         BOM=100.-BO
       WRITE(3) IL,CREPA,DEFREDO,BOM
12     PRINT 13,IL,CREPA,DEFREDO,BOM
13     FORMAT(* LEVEL N.=*,I5,* DELTA=*,G12.5,* FOR *,I5  ,* DEGREES OF F
      1REEDOM WITH *,F6.1,*  CHI-SQUARE PROBABILITY*)
       CREPA=ABS(PANF)
         DEFREDO=NOGRU
         IL=NLEV
       CALL REPRO(BO,DEFREDO,CREPA)
         BOM=100.-BO
       WRITE(3) IL,CREPA,DEFREDO,BOM
       PRINT 13,IL,CREPA,DEFREDO,BOM
10     CONTINUE
       CALL SORLEV(NLEV,NS)
       RETURN
       END
       SUBROUTINE OUS(SHAME,PANF,KI,NESDT,NQ,NOGRU,TGMED)
       DIMENSION SHAME(1),NESDT(1)
       PANF=0.
         IF(KI.EQ.1)LP=1
         NGRP=-NESDT(LP)
         I=1
       NOGRU=NGRP
         IE=1
         IIE=NESDT(LP+1)-1
2      FLE=IIE-IE+1
       AEL=0.
       DO 1 J=IE,IIE
1      AEL=AEL+SHAME(J)
       AELM=AEL/FLE
         PANF=PANF+AEL*ALOG(AELM/TGMED)
         LP=LP+1
       I=I+1
         IF(I.GT.(NGRP+1)) GO TO 3
         IE=NESDT(LP)
       IIE=NESDT(LP+1)-1
         IF(I.EQ.(NGRP+1))IIE=NQ
         GO TO 2
3      CONTINUE
       RETURN
       END
```

```
      SUBROUTINE DUINEST(NS,NQ,LB)
      COMMON NESDATA(2000),DATAR(1000),CICCA(1000),INF
      FNQ=FLOAT(NQ)
        REWIND 3
      READ(5,1) NLEV
1     FORMAT(16I5)
                              IP=0
        NLEV1=NLEV-1
        PRINT3,NLEV
3     FORMAT(*0NUMBER OF LEVELS=*,I5,* 2I NESTING*)
      IF(NLEV1.LE.0) GO TO 1000
      DO 2 I=1,NLEV1
      IUHU=I+1
      READ(5,1)NGR
        PRINT4,NGR,IUHU
4     FORMAT(1X,I5,* GROUPS IN LEVEL N.  *,I5)
      IP=IP+1
        NGR1=NGR-1
        NESDATA(IP)=-NGR1
        IP1=IP+1
      IP=IP+NGR1
        IF(IP.GT.2000)GO TO 2
      READ(5,1)(NESDATA(K),K=IP1,IP)
      PRINT5,(NESDATA(K),K=IP1,IP)
5     FORMAT(* FIRST OBJECT IN GROUPS,BESIDES N.  1*,20I5)
2     CONTINUE
1000  REWIND LB
      DO 10 I=1,NS
      PRINT 200,I
200   FORMAT(*0 SPECIES N.=*,I5)
      READ(LB) (DATAR(J),J=1,NQ)
        X1=0.
        ILV=1
      DO 11 JJ=1,NQ
      CICCA(JJ)=DATAR(JJ)
        IF(DATAR(JJ).NE.0.) X1=X1+1.
11    CONTINUE
      X2=FNQ-X1
        IF(X1.LE.0.) X1=1.E-10
       IF(X2.LE.0.)X2=1.E-10
      DUEI=2.*(-X1*ALOG(X1)-X2*ALOG(X2)+FNQ*ALOG(FNQ))
        LIBG=NQ-1
      DUEI=ABS(DUEI)
      CALL REPRO(BO,LIBG,DUEI)
        BOM=100.-BO
```

```
      WRITE(3) ILV,DUEI,LIBG,BOM
      PRINT12,ILV,DUEI,LIBG,BOM
12    FORMAT(* LEVEL N.=*,I5,*  2I=*,G12.5,*  FOR  *,I5,* DEGREES OF FRE
     1EDOM WITH *,F6.1,* CHI-SQUARE PROBABILITY*)
      IF(NLEV1.LE.0) GO TO 10
      DO 13 K=1,NLEV1
      ILV=K+1
        KI=K
        CALL IFUGO(CICCA,RELLI,KI,NESDATA,NQ,LIBG)
      DUEI=ABS(RELLI)
        CALL REPRO(BO,LIBG,DUEI)
        BOM=100.-BO
      WRITE(3) ILV,DUEI,LIBG,BOM
13    PRINT 12,ILV,DUEI,LIBG,BOM
10    CONTINUE
      CALL SORLEV(NLEV,NS)
      RETURN
      END
      SUBROUTINE IFUGO(CICCA,RELLI,KI,NESDT,NQ,LIB)
      DIMENSION CICCA(1),NESDT(1)
      RELLI=0.
        IF(KI.EQ.1)LP=1
        NGRP=-NESDT(LP)
        I=1
      COLINF=0.
        RIGINF=0.
        DETINF=0.
        TRIG1=0.
        TRIG2=0.
      IE=1
        IEE=NESDT(LP+1)-1
2     FLE=IEE-IE+1
      X1=X2=0.
      DO 1 J=IE,IEE
      IF(CICCA(J).NE.0.) X1=X1+1.
1     CONTINUE
      X2=FLE-X1
        IF(X1.LE.0.)X1=1.E-10
        IF(X2.LE.0.)X2=1.E-10
      COLINF=COLINF+FLE*ALOG(FLE)
      DETINF=DETINF+X1*ALOG(X1)+X2*ALOG(X2)
      TRIG1=TRIG1+X1
        TRIG2=TRIG2+X2
        LP=LP+1
        I=I+1
      IF(I.GT.(NGRP+1))GO TO 3
```

```
      IE=NESDT(LP)
      IEE=NESDT(LP+1)-1
      IF(I.EQ.(NGRP+1))IEE=NQ
        GO TO 2
3     LIB=NGRP
        RIGINF=TRIG1*ALOG(TRIG1)+TRIG2*ALOG(TRIG2)
      RELLI=2.*(DETINF-RIGINF-COLINF+(TRIG1+TRIG2)*ALOG(TRIG1+TRIG2))
      RETURN
      END
      SUBROUTINE REPRO(BO,INC,ALLITO)
C---IT PRODUCES NULL-HYPOTHESIS PROBABILITY FOR A GIVEN CHI-SQUARE VALUE
      DIMENSION FI(1001)
      N=INC
        COVAL=ALLITO
        NPAS=1000
      IF(N.EQ.2) GO TO 1000
      PIGR=3.141592654
      A=0.
        FN=FLOAT(N)
        RADN=SQRT(4.*FN)
        AB=FN-7.*RADN
      IF(AB.GT.A) A=AB
        IF(N.EQ.1) A=0.0642
        QPAS=FLOAT(NPAS)
      B=FN+7.*RADN
       H=(B-A)/QPAS
       FI(1)=0.
       IF(N.EQ.1)FI(1)=0.1999
      X=A
        GAML=GAM(N)
        TI=F(X,FN,GAML)
      IF(COVAL.LT.A.AND.N.NE.1) GO TO 7
        IF(COVAL.GT.B)GO TO 8
      DO 1 I=1,NPAS
      X=X+H
      TII=F(X,FN,GAML)
      DIF=(TII+TI)/2.*H
      TI=TII
1     FI(I+1)=FI(I)+DIF
      IF(N.GT.2) GO TO 2
      IF(COVAL.GT.0.0642) GO TO 2
      BO=(1.-SQRT(2.*COVAL/PIGR))*100.
      GO TO 3
2     DO 4 I=1,NPAS
      XI=A+H*FLOAT(I-1)
        XII=XI+H
```

```
      IF(COVAL.GE.XI.AND.COVAL.LE.XII) GO TO 5
4     CONTINUE
      GO TO 3
7     BO=100.
        GO TO 3
8     BO=0.
        GO TO 3
5     DFI=FI(I+1)-FI(I)
        BO=(1.-(DFI*(COVAL-XI)/H+FI(I)))*100.
3     CONTINUE
      RETURN
1000  BO=(EXP(-COVAL/2.))*100.
      RETURN
      END
      FUNCTION F(X,FN,GAML)
      F=0.
      IF(X.LE.0.) RETURN
      ALOGF=-0.6931471806+(FN/2.-1.)*ALOG(X/2.)-X/2.-GAML
      F=EXP(ALOGF)
      RETURN
      END
      FUNCTION GAM(N)
      RAPIGR=1.144729886/2.
        ILIM=N/2
      IF(MOD(N,2).EQ.0) GO TO 1
      GAM=RAPIGR
      IF(N.EQ.1) RETURN
      GAMLN=0.
      DO 2 I=1,ILIM
2     GAMLN=GAMLN+ALOG(FLOAT(I)-0.5)
      GAM=GAMLN+RAPIGR
      RETURN
1     GAMLN=0.
      ILIM=ILIM-1
      DO 3 I=1,ILIM
3     GAMLN=GAMLN+ALOG(FLOAT(I))
      GAM=GAMLN
      RETURN
      END
      SUBROUTINE SORLEV(NLEV,NS)
      COMMON NESDATA(2000),DATAR(1000),SPECR(1000),INF
      DO 100 I=1,NLEV
      PROTOT=0.
      REWIND 3
        J=0
        REWIND 4
```

```
        PRINT 200,I
200     FORMAT(*0SORTED LEVEL N.*,I5/*0  INF.VALUE-SPECIES N.-X2PROBAB.*)
1       READ(3) ILV,XDUE,NF,PROX
        IF(EOF(3)) 2,3
3       IF(ILV.NE.I) GO TO 1
        WRITE(4) ILV,XDUE,NF,PROX
        J=J+1
          DATAR(J)=PROX
        GO TO 1
2       ILM=0
5       PROMX=-500.
        DO 4 J=1,NS
        IF(DATAR(J).LE.PROMX) GO TO 4
        PROMX=DATAR(J)
          JMAX=J
4       CONTINUE
        ILM=ILM+1
          NESDATA(ILM)=JMAX
          DATAR(JMAX)=-1000.
        IF(ILM.LT.NS) GO TO 5
        DO 6 J=1,NS
        REWIND 4
          KJ=0
          ILM=NESDATA(J)
7       READ(4) ILV,XDUE,NF,PROX
          KJ=KJ+1
        IF(KJ.NE.ILM) GO TO 7
        PRINT 300,XDUE,ILM,PROX
300     FORMAT(1X,G12.5,I8 ,F10.1)
        PROTOT=PROTOT+PROX
6       CONTINUE
        PRINT 400,PROTOT
400     FORMAT("0SUM OF PROBABILITIES=",G11.4)
100     CONTINUE
        RETURN
        END

CIAHOPA
        OVERLAY(IAHOPA,0,0)
        PROGRAM MASTER(INPUT=100,OUTPUT=100,PUNCH=100,TAPE5=INPUT,TAPE6=OU
       1TPUT,TAPE7=PUNCH,TAPE1=514,TAPE2=514,TAPE3=514,TAPE4=514,TAPE8=514
       2)
        COMMON/COMUNE/INFILE    ,IFCUT,LIN,LOUT,LET
        DIMENSION LITIT(4)
C    INFILE=INPUT FILE FOR THE CHOSEN BRANCH. NOT NECESSARY FOR SCRAPS
C    IFCUT=IF GT.0,SCRAPS CUTS OUTPUT FROM MLTAX1 INTO SEGMENTS SUITABLE
```

```
C         FOR INPUT IN MLTAX2 AND MLTAX3
C   LIN,LOUT=INPUT AND OUTPUT FILE NUMBERS FOR SCRAPS
      WRITE(6,100)
100   FORMAT(1HT)
1     READ(5,2)(LITIT(I),I=1,4)
2     FORMAT(4A10)
3     READ(5,5) MLTAX,INFILE,IFCUT,LIN,LOUT
5     FORMAT(A10,4I5)
      IF(EOF(5)) 4,9
9     WRITE(6,6)(LITIT(I),I=1,4)
6     FORMAT(1H1,4A10,*   IAHOPA*)
      WRITE(6,7) MLTAX,INFILE,IFCUT,LIN,LOUT
7     FORMAT(* REQUIRED PROGRAM ,INPUT FILE ,IFCUT,LIN,LOUT*/1X,A10,4I5,
     1* IAHOPA*)
      IF(MLTAX   .EQ.(6HMLTAX1)) CALL OVERLAY(6HIAHOPA,1,0)
      IF(MLTAX.EQ.(6HSCRAPS)) CALL OVERLAY(6HIAHOPA,4,0)
      IF(MLTAX   .EQ.(6HMLTAX2)) CALL OVERLAY(6HIAHOPA,2,0)
      IF(MLTAX   .EQ.(6HMLTAX3)) CALL OVERLAY(6HIAHOPA,3,0)
      GO TO 3
4     REWIND1
        REWIND4
        REWIND8
      REWIND2
        REWIND 3
      WRITE(6,8)
8     FORMAT(*0IAHOPA TERMINATED*)
      STOP
      END
      OVERLAY(1,0)
      PROGRAM MLTAX1
      DIMENSION ITITLE(4),NOUN(4),LORDER(118),ISPOC(118)
      COMMON/COMUNE/INFILE,IFCUT,LIN,LOUT,LET
      COMMON DATA(11800),SIM(118,118),IBUF(118),ISPEC(118),ILCOPI(118),
     1INP,IZIP
C
C   DATA INPUT FILES AVAILABLE=5,3,8(NO 3 IF DATA FOR SCRAPS PRODUCED)
C   DATA=DATA MATRIX,WITH DATA PACKED 10 PER WORD
C   SIM=SIMILARITY MATRIX
C   NSPEC=NUMBER OF SPECIES(UP TO 118)
C   NQUAD=NUMBER OF RELEVES(UP TO 1000 IF ITRASP GT.0,UP TO 250 OTHER
C       WISE)
C   ITRASP=IF EQ.0,DATA ENTERED BY SPECIES,IF GT.0,BY RELEVES
C   IFPR=IF GT.0,SIMILARITY MATRIX PRINTED
C   TITLE=LABEL OF ACTUAL RUN
C   NOUN=ALPHANUMERIC NAME OF EACH SPECIES
C   ISPEC=VECTOR OF NUMERICAL LABELS CORRESPONDING TO SPECIES
```

```
C     NAMES,USED FOR ORDERING
C   LASOLP=IF GT.0,MATRIX SIM IS COMPUTED,USING PRESENCE/ABSENCE SCORES
C   NUDATA=NUMBER OF SPECIES IN RELEVES(ITRASP MUST BE GT.0). MAY BE
C     UP TO 250,BUT ONLY 118 CAN BE CHOSEN. NUDATA=NQUAD IF ITRASP
C   EQ.0
C   INSEQ=IF GT.0,ORDERING FOLLOWS INSEQ(EXTERNAL SEQUENCES)
C         NEITHER SIM NOR DENDROGRAM TABLE PRODUCED. A VECTOR OF
C         ORDERED RELEVES IS WRITTEN IN FILE 4 FOR EACH SEQUENCE
C   ISIM=IF GT.0,SIMILARITY MATRIX(LOWER TRIANGLE) WRITTEN IN FILE 4
C     (10X,10G12.5)
C   IFULL=IF GT.0,RELEVES ORDERED BASED ON ACTUAL DATA,ELSE
C         ON PRESENCE/ABSENCE SCORES
C   IRAJSKI=IF 0,SIMILARITY MATRIX BY JACCARD INDEX,IF GT.0  BY RAJSKI
C          INDEX (EQUIVOCATION/JOINT)
C   LORDER=EXTERNAL ORDERING SEQUENCE(ONLY IF INSEQ.GT.0)
C   ISPOC=VECTOR OF ORIGINAL SPECIES LABEL(USED ONLY
C         IF INSEQ GT.1)
C   ILCOPI=A CODE OF 7 DIGITS FOR SPECIES
C
      REWIND 4
      REWIND 3
      INP=INFILE
        IZIP=IFCUT
        NUT=0
9     READ(INP,1) (ITITLE(I),I=1,4)
    1 FORMAT(4A10)
      IF(ITITLE(1).EQ.10HFINEMLTAX1) GO TO 1000
11    READ(INP,2)NSPEC,NQUAD,NUDATA,ITRASP,IFPR,LASOLP,INSEQ,ISIM,IFULL
     1,IRAJSKI
      NUT=NUT+1
    2 FORMAT(16I5)
      WRITE(6,10)(ITITLE(I),I=1,4),NSPEC,NQUAD,NUDATA,ITRASP,IFPR,LASOLP
     1,INSEQ,ISIM,IFULL,IRAJSKI
10    FORMAT(1H1,4A10/* MLTAX1-NS NQ ND ITR IPR LSLP INSQ ISM IFL IRAJSK
     1I*/1X,10I5)
      IF(IZIP.GT.0) WRITE(LIN,12) NUT,NSPEC,NQUAD,(ITITLE(I),I=1,4)
C
C   READS AND PRINTS PAIRS OF NUMERIC AND ALPHANUMERIC LABELS
C
      DO 3 I=1,NSPEC
      READ(INP,4) ISPEC(I),ILCOPI(I),(NOUN(K),K=1,4)
4     FORMAT(I3,I7,4A10)
      ISPOC(I)=ISPEC(I)
3     WRITE(6,12) I,ISPEC(I),ILCOPI(I),(NOUN(K),K=1,4)
12    FORMAT(2X,3I8,2X,4A10)
C
```

```
C    READS RELEVE DATA IN SUBROUTINE SQUEEZ
C
     CALL SQUEEZ(NSPEC,NQUAD,NUDATA,ITRASP)
     IF(INSEQ.GT.0) GO TO 5
C
C    COMPUTES SIMILARITY MATRIX
C
     IF(IRAJSKI.EQ.0)GO TO 1003
     CALL RAJSKI(NSPEC,NQUAD)
     GO TO1004
1003 CALL CPSIM(NSPEC,NQUAD,LASOLP)
1004 IF(IFPR.EQ.0) GO TO 5
C
C   PRINTS ROWS OF SIMILARITY MATRIX(IF REQUESTED)
C
     WRITE(6,1002)
1002 FORMAT(1X////1X,* SIMILARITY MATRIX-   MLTAX1*)
     DO 6 K=2,NSPEC
     WRITE(6,8) K
   8 FORMAT(1X,5HROW =,I4)
     I1=K-1
     IF(ISIM.GT.0) WRITE(4,7)(SIM(K,I),I=1,I1)
6    WRITE(6,7) (SIM(K,I),I=1,I1)
7    FORMAT(10X,10G12.5)
C
C   PERFORMS ORDERING OF SPECIES FROM SIMILARITY MATRIX,IF INSEQ.LE.0,
C  OR FROM EXTERNAL INPUT
C
5    DO 13 I=1,INSEQ
     IF(I.LT.2) GO TO 14
     DO 15 JM=1,NSPEC
15   ISPEC(JM)=ISPOC(JM)
14   IF(INSEQ.GT.0) READ(INP,2) (LORDER(MJ),MJ=1,NSPEC)
     CALL ORDSPEC(NSPEC,INSEQ,LORDER,LIN)
C
C   PERFORMS ORDERING OF RELEVES ACCORDING TO SPECIES
C    PRESENCE OR QUANTITY
13    CALL ORDRELV(NSPEC,NQUAD,LIN,IFULL,INSEQ)
     GO TO 9
1000  WRITE(6,1001) ITITLE(1)
1001  FORMAT(1X////*  WORK REGULARLY TERMINATED-  MLTAX1*,A10)
     LET=NUT
     END
     SUBROUTINE SQUEEZ(NSPEC,NQUAD,NUM,ITRASP)
     COMMON DATA(11800),SIM(118,118),IBUF(118),ISPEC(118),ILCOPI(118),
    1INP,IZIP
```

```
C    IF=INPUT FORMAT FOR DATA TABLES,
C    IND=THEIR INPUT FILE-IF NOT SPECIFIED,IND=INP
      DIMENSION IF(7)
      REAL LSTR(250)
      READ(INP,100) IF,IND
100   FORMAT(7A10,5X,I5)
      IF(IND.LE.0) IND=INP
      PRINT 101,IF,IND
101   FORMAT(1X,7A10,I5)
      WRITE(6,400)
400   FORMAT(1X////*  DATA TABLE-  MLTAX1*)
      K=1
      IF(ITRASP.EQ.0) GO TO 7
      DO 1 I=1,NQUAD
      WRITE(6,500) I
500   FORMAT(*  DATA ROW N,*,I5)
C
C    READS UP TO NQUAD RELEVES
C
      READ(IND,IF)(LSTR(N),N=1,NUM)
C
C    PACKS RELEVE DATA
C
      DO 6 N=1,NSPEC
      MM=ISPEC(N)
6     IBUF(N)=LSTR(MM)
      WRITE(6,300)(IBUF(L),L=1,NSPEC)
300   FORMAT(1X,40I3)
      DO 2 J=1,NSPEC
4     CALL MOVCHR(10,IBUF(J),K,DATA)
      K=K+1
      IF(K.GT.118000) GO TO 3
    2 CONTINUE
    1 CONTINUE
      RETURN
C
C    READS NSPEC ROWS OF NQUAD RELEVES
C
7     DO 8 I=1,NSPEC
      WRITE(6,500) I
      READ(IND,IF)(LSTR(N),N=1,NQUAD)
      WRITE(6,301)(LSTR(N),N=1,NQUAD)
301   FORMAT(1X,30F4.0)
      DO 9 J=1,NQUAD
      INPOS=I+(J-1)*NSPEC
      LASTRA=LSTR(J)
```

```
      CALL MOVCHR(10,LASTRA,INPOS,DATA)
      K=K+1
      IF(K.GT.118000) GO TO 3
9     CONTINUE
8     CONTINUE
      RETURN
C
C  MESSAGE ABOUT BAD DATA
C
   3 WRITE(6,200) I,J
200    FORMAT(*  TOO MUCH DATA IN MLTAX1-RELEN.*,I5,*SPN.*,I5)
      RETURN
      END
      SUBROUTINE CPSIM(NSPEC,NQUAD,IE)
      COMMON DATA(11800),SIM(118,118),IBUF(118),ISPEC(118),ILCOPI(118),
     1INP,IZIP
      DO 3 I=1,NSPEC
      DO 3 K=1,NSPEC
3     SIM(I,K)=0.
      DO 1 I=1,NSPEC
      DO 1 K=1,NSPEC
      IF(I.GE.K) GO TO 1
      PRSC=0.
      PNORMI=0.
      PNORMK=0.
      DO 2 J=1,NQUAD
      NIJ=(J-1)*NSPEC+I
      NJK=(J-1)*NSPEC+K
      IPARK=0
      IPORK=0
      CALL MOVCHR(NIJ,DATA,10,IPARK)
      CALL MOVCHR(NJK,DATA,10,IPORK)
      IF(IE.LE.0) GO TO 4
      IF(IPARK.GT.0) IPARK=1
      IF(IPORK.GT.0) IPORK=1
4     DPRSC=IPARK*IPORK
      DNORMI=IPARK**2
      DNORMK=IPORK**2
      PRSC=PRSC+DPRSC
      PNORMI=PNORMI+DNORMI
   2 PNORMK=PNORMK+DNORMK
      DENOM=PNORMI+PNORMK-PRSC
      IF(DENOM.LE.0.) DENOM=-1.E-6
      SIM(I,K)=PRSC/DENOM
      SIM(K,I)=SIM(I,K)
   1 CONTINUE
```

104

```
      RETURN
      END
      SUBROUTINE RAJSKI(NSPEC,NQUAD)
      COMMON DATA(11800),SIM(118,118),IBUF(118),ISPEC(118),ILCOPI(118),
     1INP,IZIP
      DO 3 I=1,NSPEC
      DO 3 K=1,NSPEC
3     SIM(I,K)=0.
      DO 1 I=1,NSPEC
      DO 1 K=1,NSPEC
      IF(I.GE.K)GO TO 1
      TRIG=TCOL1=TCOL2=SJK=SRIG=SCOL=0.
      DO 2 J=1,NQUAD
      NIJ=(J-1)*NSPEC+I
       NJK=(J-1)*NSPEC+K
      IMAT=KMAT=0
      CALL MOVCHR(NIJ,DATA,10,IMAT)
       CALL MOVCHR(NJK,DATA,10,KMAT)
      AMAT=IMAT
       IF(AMAT.LE.0.)AMAT=1.E-9
       BMAT=KMAT
      IF(BMAT.LE.0.)BMAT=1.E-9
       TRIG=AMAT+BMAT
      TCOL1=TCOL1+AMAT
       TCOL2=TCOL2+BMAT
      SJK=-(AMAT*ALOG(AMAT)+BMAT*ALOG(BMAT))+SJK
2     SRIG=-TRIG*ALOG(TRIG)+SRIG
      SCOL=-TCOL1*ALOG(TCOL1)-TCOL2*ALOG(TCOL2)
      TOG=TCOL1+TCOL2
      GIOMU=ABS(2.*SJK-SRIG-SCOL)
       SJK=ABS(SJK+TOG*ALOG(TOG))
      SIM(I,K)=GIOMU/SJK
       SIM(K,I)=SIM(I,K)
1     CONTINUE
      RETURN
      END
      SUBROUTINE ORDSPEC(NSPEC,LAPEP,IZOZZER,LIN)
      COMMON DATA(11800),SIM(118,118),IBUF(118),ISPEC(118),ILCOPI(118),
     1INP,IZIP
      DIMENSION IORD(118),IZOZZER(1)
      NPROG=0
      IF(LAPEP.LE.0) GO TO 14
      DO 15 I=1,NSPEC
15    IBUF(I)=IZOZZER(I)
      GO TO 11
14    IACEGAT=0
```

```
      IMAX=0
      KMAX=0
      SMAX=-1.E-6
      IEFATTD=NSPEC*(NSPEC-1)/2
      IORD(1)=10H.                  .
      DO 12 I=2,10
12    IORD(I)=10H           .
      WRITE(6,200)
200   FORMAT(1X////4X,16HDENDROGRAM TABLE,9X,1H1,8X,2H.9,8X,2H.8,8X,2H.7
     1,8X,2H.6,8X,2H.5,8X,2H.4,8X,2H.3,8X,2H.2,8X,2H.1,8X,2H.0/1X)
      DO 6 I=1,NSPEC
6     IBUF(I)=0
C
C  FINDS FIRST MAXIMUM
C
      DO 1 I=1,NSPEC
      DO 1 K=I,NSPEC
      IF(SIM(I,K).LE.SMAX) GO TO 1
      SMAX=SIM(I,K)
      IMAX=I
      KMAX=K
1     CONTINUE
      IBUF(1)=IMAX
      IBUF(2)=KMAX
      IFOUND=1
      IACEGAT=IACEGAT+1
      VALMAX=SIM(IMAX,KMAX)
      WRITE(6,100) IMAX,KMAX,VALMAX,ISPEC(IMAX),ISPEC(KMAX),(IORD(IF),IF
     1=1,10)
100   FORMAT(1X,2I4,G12.5,2I4,10A10)
      SIM(IMAX,KMAX)=-IACEGAT
      SIM(KMAX,IMAX)=-IACEGAT
C
C  SCANS THE ROWS OF ALREADY FOUND MAXIMA TO FIND
C    THE NEXT MAXIMUM
5     SMAX=-1.E-6
      IMAX=0
      KMAX=0
      DO 2 M=1,NSPEC
      IF(IBUF(M).LE.0) GO TO 3
      IRIG=IBUF(M)
      DO 2 K=1,NSPEC
      IF(SIM(IRIG,K).LE.SMAX) GO TO 2
      SMAX=SIM(IRIG,K)
      IMAX=IRIG
      KMAX=K
```

```
2      CONTINUE
3      VALMAX=SIM(IMAX,KMAX)
       IACEGAT=IACEGAT+1
       SIM(KMAX,IMAX)=-IACEGAT
       SIM(IMAX,KMAX)=-IACEGAT
       IF(IACEGAT.GE.IEFATTO) GO TO 11
       JU=0
        JE=0
       DO 4 J=1,NSPEC
       IF(IBUF(J).EQ.IMAX) JU=1
       IF(IBUF(J).EQ.KMAX) JE=1
4      CONTINUE
       IF(JU.EQ.1.AND.JE.EQ.1) GO TO 8
C
C   STORES NEWLY FOUND MAXIMUM
C
       DO 7 M=1,NSPEC
       IF(IBUF(M).GT.0) GO TO 7
       IF(JU.EQ.1) IBUF(M)=KMAX
       IF(JE.EQ.1) IBUF(M)=IMAX
       IFOUND=IFOUND+1
       IF(MOD(IFOUND,10).EQ.0) GO TO 13
       WRITE(6,100) IMAX,KMAX,VALMAX,ISPEC(IMAX),ISPEC(KMAX)
       GO TO 8
13     WRITE(6,100) IMAX,KMAX,VALMAX,ISPEC(IMAX),ISPEC(KMAX),(IORD(IP),IP
      1=1,10)
       GO TO 8
7      CONTINUE
8      IF(IBUF(NSPEC).EQ.0) GO TO 5
C
C   PRINTS HEADLINE
C
11     DO 9 I=1,NSPEC
       IUK=IBUF(I)
9      IORD(I)=ISPEC(IUK)
       WRITE(6,300)
300    FORMAT(1X///////////////////1X)
       CALL PRIRI(IORD,-NSPEC,0,NPROG)
       DO 10 I=1,NSPEC
       ISPEC(I)=IBUF(I)
       IUK=IBUF(I)
       IORD(I)=ILCOPI(IUK)
10     IBUF(I)=0
       IF(IZIP.GT.0) WRITE(LIN,400) (IORD(J),J=1,NSPEC)
400    FORMAT(10I8)
       CALL PRIRI(IORD,-NSPEC,1,NPROG)
```

```
      RETURN
      END
      SUBROUTINE ORDRELV(NSPEC,NQUAD,LIN,MABASTA,INSEQ)
      COMMON DATA(11800),SIM(118,118),IBUF(118),ISPEC(118),ILCOPI(118),
     1INP,IZIP
       DIMENSION IOI(118),RELSEQ(1000),LOPAIO(13),MASSI(12)
      NPROG=0
      ILP=1
      ILS=2
      DO 18 KK=1,13
18    MASSI(KK)=0
      IPA=0
      IIPA=0
      KMAX=1
      IMAX=1
      REWIND ILP
      I=1
1     DO 19 KK=1,12
19    LOPAIO(KK)=0
      LOPAIO(13)=I
      J=1
2     CALL MOVCHR((I-1)*NSPEC+J,DATA,10,IBUF(J))
      J=J+1
      IF(J.LE.NSPEC) GO TO 2
      J=1
3     MIAU=ISPEC(J)
      IF(IBUF(MIAU).EQ.0) GO TO 4
        LUNO=IBUF(MIAU)
        IF(MABASTA.LE.0)    LUNO=1
      CALL MOVCHR(10,LUNO,J,LOPAIO)
4     IOI(J)=IBUF(MIAU)
      J=J+1
      IF(J.LE.NSPEC) GO TO 3
      DO 100 KK=1,NSPEC
      CALL MOVCHR(KK,LOPAIO,10,IPA)
      CALL MOVCHR(KK,MASSI,10,IIPA)
      IF(IPA-IIPA)8,100,7
100   CONTINUE
      GO TO 8
7     DO 20 KK=1,12
20    MASSI(KK)=LOPAIO(KK)
      KMAX=I
8     WRITE(ILP) LOPAIO,(IOI(M),M=1,NSPEC)
      I=I+1
      IF(I.LE.NQUAD) GO TO 1
      J=1
```

```
       NREC=NQUAD
9      DO 21 KK=1,12
21     MASSI(KK)=0
       REWIND ILP
       REWIND ILS
       I=1
   10 READ(ILP) LOPAIO,(IOI(M),M=1,NSPEC)
       LOP3=LOPAIO(13)
       IF(LOP3.NE.KMAX) GO TO 11
       IF(IZIP.GT.0) WRITE(LIN,16)(IOI(M),M=1,NSPEC)
16     FORMAT(40I2)
       RELSEQ(J)=LOP3
       CALL PRIRI(IOI,NSPEC,LOP3,NPROG)
       GO TO 13
   11 DO 101 KK=1,NSPEC
       CALL MOVCHR(KK,LOPAIO,10,IPA)
       CALL MOVCHR(KK,MASSI,10,IIPA)
       IF(IPA-IIPA)15,101,14
101    CONTINUE
       GO TO 15
14     DO 22 KK=1,12
   22 MASSI(KK)=LOPAIO(KK)
       IMAX=LOPAIO(13)
   15 WRITE(ILS) LOPAIO,(IOI(M),M=1,NSPEC)
   13 I=I+1
       IF(I.LE.NREC) GO TO 10
       NREC=NREC-1
       KMAX=IMAX
       J=J+1
       IF(J.GT.NQUAD) GO TO 12
       LUI=ILP
       ILP=ILS
       ILS=LUI
       GO TO 9
12     IF(INSEQ.GT.0) WRITE(4,17)(RELSEQ(MUS),MUS=1,NQUAD)
17     FORMAT(10F8.0)
       RETURN
       END
       SUBROUTINE PRIRI(IRIF,ISW,IQ,NPROG)
       DIMENSION IRIGA(12),IRIF(1)
       MAX=0
       IBLANK=10H
       IPUNTO=10H
       LABELX=10H SPECIES
       LABELY=10H RELEVES
       DO 1 I=1,12
```

```
      1 IRIGA(I)=IBLANK
        IF(ISW.GT.0) GO TO 2
        ISW=-ISW
        IDIV=1000000
        DO 7 K=1,7
        DO 3 I=1,ISW
        IF(K.EQ.7) GO TO 4
        IACI=IRIF(I)/IDIV
        IRIF(I)=IRIF(I)-IACI*IDIV
        IF(IACI.LE.0.OR.IACI.GE.10) GO TO 5
        IACI=IACI+27
        GO TO 6
      5 IACI=33B
        IF(IRIF(I).GT.0) IACI=55B
6       CALL MOVCHR(10,IACI,I,IRIGA)
        GO TO 3
      4 IACI=MOD(IRIF(I),10)
        IACI=IACI+27
        GO TO 6
      3 CONTINUE
        IDIV=IDIV/10
        LAPEL=IBLANK
        IF(IQ.NE.0) GO TO 7
        IF(K.EQ.5) LAPEL=LABELX
7       WRITE(6,8) LAPEL,(IRIGA(L),L=1,12)
8       FORMAT(2X,13A10)
        IF(IQ.NE.0) WRITE(6,8) LABELY
        RETURN
      2 DO 9 I=1,ISW
        ICAR=IRIF(I)+27
        IF(IRIF(I).GT.MAX) MAX=IRIF(I)
        IF(ICAR.EQ.33B) ICAR=IPUNTO
9       CALL MOVCHR(10,ICAR,I,IRIGA)
        NPROG=NPROG+1
        PRINT 13,NPROG,IQ
13      FORMAT(1X,2I5)
        DO 11 MAP=1,MAX
        PRINT 10,(IRIGA(L),L=1,12)
10      FORMAT(1H+,11X,12A10)
        DO 11 IA=1,ISW
        IF(IRIF(IA).LE.MAP) CALL MOVCHR(10,IBLANK,IA,IRIGA)
11      CONTINUE
        RETURN
        END
        OVERLAY(2,0)
        PROGRAM MLTAX2
```

```
      COMMON/COMUNE/INFILE    ,IFCUT,LIN,LOUT,LET
      COMMON DATA(3750),ICOOC(11325),IBUF(250),ISPEC(150),ISOL(150),TMAR
     1G(150),INF
      DIMENSION ITITLE(4),NOUN(4),ILCOPI(150)
C
C    DATA INPUT FILES AVAILABLE=5,3,8(3 IF DATA FROM SCRAPS)
C    DATA=DATA MATRIX(10 DATA PER WORD)
C    IF ITRASP EQ.0,DATA ENTERED BY SPECIES.BUT THEN NUDATA=NQ.IF ITRASP
C    GT.0,DATA ENTERED BY RELEVES
C    ICOOC=CO-OCCURRENCE MATRIX
C    NSPEC=UP TO 150 SPECIES
C    NQUAD=UP TO 250 RELEVES
C    IFPR=IF GT.0,ICOOC AND MARGINAL TOTALS OF INPUT DATA(WHEN IENT GT.0)
C     ARE PRINTED
C    ITITLE=LABEL OF ACTUAL RUN
C    NOUN=ALPHANUMERIC LABEL OF SPECIES
C    ISPEC=NUMERICAL LABEL OF SPECIES
C    NUDATA=NUMBER OF DATA IN RELEVES(ITRASP=1)
C    IENT=IF GT.0,ENTROPY OF INPUT DATA COMPUTED
C    NEWDIAG=IF GT.0, CO-OCCURRENCE MATRIX DIAGONAL ELEMENTS REPLACED BY
C          SPECIES FREQUENCY IN THE SAMPLE
C
      INF=INFILE
9     READ(INF,1) (ITITLE(I),I=1,4)
    1 FORMAT(4A10)
      IF(ITITLE(1).EQ.10HFINEMLTAX2) GO TO 1000
      IF(ITITLE(1).EQ.10HSCRAPS-MLT) GO TO 1000
11    READ(INF,2)NSPEC,NQUAD,NUDATA,ITRASP,IFPR,IENT,NEWDIAG
    2 FORMAT(16I5)
      WRITE(6,10) (ITITLE(I),I=1,4),NSPEC,NQUAD,IFPR,NUDATA,ITRASP,IENT
     1,NEWDIAG
10    FORMAT(1H1,4A10/* MLTAX2-NS NQ IPR NDT ITR IH NEWDIAG*/1X,7I5)
      DO 3 I=1,NSPEC
      READ(INF,4) ISPEC(I),ILCOPI(I),(NOUN(K),K=1,4)
4     FORMAT(I3,I7,4A10)
3     WRITE(6,12) I,ISPEC(I),ILCOPI(I),(NOUN(K),K=1,4)
12    FORMAT(2X,3I8,2X,4A10)
      CALL LFTR(NSPEC,NQUAD,NUDATA,ITAP,ITRASP)
      CALL STRUK(NSPEC,NQUAD,ITAP,NMED,IENT,IFPR)
      CALL CPCOOC(NSPEC,NQUAD,NEWDIAG)
      IF(IFPR.EQ.0) GO TO 5
      WRITE(6,1002)
1002  FORMAT(1X////*  CO-OCCURRENCE MATRIX-  MLTAX2*)
      DO 6 K=1,NSPEC
      WRITE(6,8) K
    8 FORMAT(1X,5HROW =,I4)
```

```
      I1=ICOL(NSPEC,K)+1
      I2=ICOL(NSPEC,K)+NSPEC-(K-1)
    6 WRITE(6,7) (ICOOC(I),I=I1,I2)
    7 FORMAT(1X,10I4,5X,10I4,5X,10I4)
    5 CALL CPENTR(ICOOC,NSPEC,NQUAD,TMARG)
      GO TO 9
 1000 IF(ITITLE(1).EQ.10HSCRAPS-MLT) READ(5,1) ITITLE(2)
      WRITE(6,1001) ITITLE(1)
 1001 FORMAT(1X////*WORK REGULARLY TERMINATED-MLTAX2*,A10)
      END
      SUBROUTINE LFTR(NSPEC,NQUAD,NUDATA,ITAP,ITRASP)
      COMMON DATA(3750),ICOOC(11325),IBUF(250),ISPEC(150),ISOL(150),TMAR
     1G(150),INF
C   IF=INPUT FORMAT FOR DATA TABLES
C   IND=THEIR INPUT FILE-IF NOT SPECIFIED,IND=INF
      DIMENSION STR(250),IF(7)
C   TO ADAPT ITRASP OPTION AS IN MLTAX3
      IF(ITRASP.EQ.0) GO TO 10
      ITRASP=-1
   10 ITRASP=ITRASP+1
      READ(INF,100) IF,IND
 100  FORMAT(7A10,5X,I5)
      IF(IND.LE.0) IND=INF
      PRINT 101,IF,IND
 101  FORMAT(1X,7A10,I5)
      ITAP=1
      REWIND 2
      REWIND 1
      IF(ITRASP.GT.0) GO TO 1
      DO 2 I=1,NQUAD
      READ(IND,IF) (STR(N),N=1,NUDATA)
      DO 3 NN=1,NSPEC
      MM=ISPEC(NN)
 3    IBUF(NN)=STR(MM)
 2    WRITE(1) (IBUF(N),N=1,NSPEC)
      REWIND 1
      RETURN
 1    DO 4 I=1,NSPEC
      READ(IND,IF) (STR(N),N=1,NUDATA)
 4    WRITE(1) (STR(N),N=1,NUDATA)
      ITAP=2
      REWIND 2
      DO 5 J=1,NQUAD
      REWIND 1
      DO 6 I=1,NSPEC
      READ(1) (STR(L),L=1,NUDATA)
```

112

```
6     IBUF(I)=STR(J)
5     WRITE(2) (IBUF(L),L=1,NSPEC)
      REWIND 2
      REWIND 1
      RETURN
      END
      SUBROUTINE STRUK(NSPEC,NQUAD,ITAP,NMED,IENT,IFPR)
      COMMON DATA(3750),ICOOC(11325),IBUF(250),ISPEC(150),ISOL(150),TMAR
     1G(150),INF
      DIMENSION QMARG(250)
      TG=0.
      DO 4 JJ=1,NSPEC
4     TMARG(JJ)=0.
      DO 5 JJ=1,NQUAD
5     QMARG(JJ)=0.
      WRITE(6,400)
400   FORMAT(1X////*  DATA TABLE-  MLTAX2*)
      NMED=0
      DO 11 JJ=1,NSPEC
   11 ISOL(JJ)=0
      K=1
      DO 1 I=1,NQUAD
      WRITE(6,500) I
500   FORMAT(* REL.*,I5)
      READ(ITAP) (IBUF(L),L=1,NSPEC)
      WRITE(6,300) (IBUF(L),L=1,NSPEC)
  300 FORMAT(6X,10I3,3X,10I3,3X,10I3,3X,10I3)
      INDIC=0
      ISUM=0
      DO 10 JJ=1,NSPEC
      QMARG(I)=QMARG(I)+IBUF(JJ)
      TMARG(JJ)=TMARG(JJ)+IBUF(JJ)
      TG=TG+IBUF(JJ)
      IDSUM=0
      IF(IBUF(JJ).LE.0) GO TO 10
      IDSUM=1
      INDIC=JJ
      ISUM=ISUM+IDSUM
   10 CONTINUE
      IF(INDIC.GT.0.AND.ISUM.EQ.1) ISOL(INDIC)=ISOL(INDIC)+1
      NMED=NMED+ISUM
      DO 2 J=1,NSPEC
      CALL MOVCHR(10,IBUF(J),K,DATA)
      K=K+1
      IF(K.GT.37500) GO TO 3
    2 CONTINUE
```

```
      1 CONTINUE
      IF(IENT.LE.0) RETURN
      WRITE(6,100)
100   FORMAT(*  CALCULATIONS ON DATA MATRIX-  MLTAX2*)
      NTOT=NQUAD*NSPEC
      HXY=0.
      DO 6 JJ=1,NTOT
      IPIRK=0
      CALL MOVCHR(JJ,DATA,10,IPIRK)
      PIK=FLOAT(IPIRK)/TG
      IF(PIK.LE.0.) PIK=1.
      DHXY=PIK*ALOG(PIK)
6     HXY=HXY+DHXY
      HY=0.
      DO 7 JJ=1,NSPEC
      PIK=TMARG(JJ)/TG
      IF(PIK.LE.0.) PIK=1.
      IF(IFPR.GT.0) WRITE(6,13) JJ,TMARG(JJ)
13    FORMAT(* SPECIES N.*,I4,5X,*MARGINAL TOT.=*,F10.0)
      DHY=PIK*ALOG(PIK)
7     HY=HY+DHY
      HX=0.
      DO 8 JJ=1,NQUAD
      PIK=QMARG(JJ)/TG
      IF(PIK.LE.0.) PIK=1.
      IF(IFPR.GT.0) WRITE(6,14) JJ,QMARG(JJ)
14    FORMAT(* QUADRAT N.*,I4,5X,*MARGINAL TOT.=*,F10.0)
      DHX=PIK*ALOG(PIK)
8     HX=HX+DHX
      H=-(HX+HY-HXY)
      HMAX=ALOG(FLOAT(NSPEC))
      RIDOND=1.-H/HMAX
      HTAB=0.
       RITAB=0.
       ENM=NSPEC
      IF((NMED/NQUAD).EQ.NSPEC) GO TO 12
      ENM=NMED/NQUAD
      HTAB=ENM*ALOG(FLOAT(NSPEC)/ENM)
      RITAB=1.-H/HTAB
12    WRITE(6,9) HX,HY,HXY,H,HMAX,RIDOND,HTAB,RITAB,ENM
9     FORMAT(* HX=*,E10.3/* HY=*,E10.3/* HXY=*,E10.3/* H=*,E10.3/* HMAX=
     1*,E10.3/* 1-H/HMAX=*,E10.3/* N  LN (NSPEC/N)=*,E10.3/* 1-H/HTAB=*,
     2E10.3/* MEAN SPEC.NUMB./QUADR.=*,F10.0)
      RETURN
    3 WRITE(6,200) I,J
200   FORMAT(*  TOO MANY DATA IN MLTAX2-RELN.*,I5,*SPN.*,I5)
```

```
      RETURN
      END
      SUBROUTINE CPCOOC(NSPEC,NQUAD,ND)
      COMMON DATA(3750),ICOOC(11325),IBUF(250),ISPEC(150),ISOL(150),TMAR
     1G(150),INF
      NPOST=NSPEC*(NSPEC+1)/2
      DO 3 I=1,NPOST
    3 ICOOC(I)=0
      DO 1 I=1,NSPEC
      DO 1 K=I,NSPEC
      IPRSC=0
      IND=ICOL(NSPEC,I)+K-(I-1)
      IF(ND.GT.0)GO TO 5
      IF(K.EQ.I) GO TO 4
    5 DO 2 J=1,NQUAD
      NIJ=(J-1)*NSPEC+I
      NJK=(J-1)*NSPEC+K
      IPARK=0
        IPORK=0
      CALL MOVCHR(NIJ,DATA,10,IPARK)
      CALL MOVCHR(NJK,DATA,10,IPORK)
      IF(IPARK.GT.0) IPARK=1
      IF(IPORK.GT.0) IPORK=1
      IDPRSC=IPARK*IPORK
    2 IPRSC=IPRSC+IDPRSC
      ICOOC(IND)=IPRSC
      GO TO 1
    4 IF(K.EQ.I) ICOOC(IND)=ISOL(I)
    1 CONTINUE
      RETURN
      END
      SUBROUTINE CPENTR(ICOOC,NSPEC,NQUAD,TMARG)
      DIMENSION ICOOC(1),TMARG(1)
      WRITE(6,100)
  100 FORMAT(*  CALCULATIONS ON CO-OCCURRENCE MATRIX-MLTAX2*)
      DO 4 I=1,NSPEC
      TMARG(I)=0.
      TINCR=0
      DO 1 K=1,NSPEC
      IF(K.LT.I) GO TO 2
      IROW=I
      KOL=K
      GO TO 3
    2 IROW=K
      KOL=I
    3 IND=ICOL(NSPEC,IROW)+KOL-(IROW-1)
```

```
      TINCR=ICOOC(IND)
    1 TMARG(I)=TMARG(I)+TINCR
    4 WRITE(6,5) I,TMARG(I)
    5 FORMAT(1X,7HROW.N.=,I5,10X,10HMARG.TOT.=,F10.0)
      TOTG=0.
      DO10 I=1,NSPEC
10    TOTG=TOTG+TMARG(I)
      WRITE(6,6) TOTG
    6 FORMAT(1X, 9HGEN.TOT.=,F10.0////1X)
      HMARG=0.
      DO 7 I=1,NSPEC
      PI=TMARG(I)/TOTG
      IF(PI.LE.0.) PI=1.
      PILOG2=PI*ALOG(PI)
    7 HMARG=HMARG+PILOG2
      HMARG=HMARG*2.
      HINT=0.
      DO 8 I=1,NSPEC
      DO 8 K=I,NSPEC
      IND=ICOL(NSPEC,I)+K-(I-1)
      PIK=ICOOC(IND)/TOTG
      IF(PIK.LE.0.) PIK=1.
      PIKLOG2=PIK*ALOG(PIK)
      IF(K.NE.I) PIKLOG2=PIKLOG2*2.
    8 HINT=HINT+PIKLOG2
      ENTR=-(HMARG-HINT)
      HMAX=ALOG(FLOAT(NSPEC))
      RIDOND=1.-(ENTR/HMAX)
      WRITE(6,9) HMARG,HINT,ENTR,HMAX,RIDOND
9     FORMAT(* HX+HY=*,E10.3/* HXY=*,E10.3/* H=*,E10.3/* HMAX=*,E10.3/*
     11-H/HMAX=*,E10.3)
      RETURN
      END
      FUNCTION ICOL(NUM,IRIG)
      ICOL=0
      IF(IRIG.EQ.1) RETURN
      LARIG=IRIG-1
      DO 1 MM=1,LARIG
    1 ICOL=ICOL+NUM-MM+1
      RETURN
      END
      OVERLAY(3,0)
      PROGRAM MLTAX3
      COMMON/COMUNE/INFILE   ,IFCUT,LIN,LOUT,LET
      COMMON DATA(3750),ICOOC(11325),IBUF(250),ISPEC(150),ISOL(150),TMAR
     1G(150),RID(60,60),INT
```

```
      DIMENSION ITITLE(4),NOUN(4),MTITLE(4),ILCOPI(150)
C
C   DATA INPUT FILES AVAILABLE=5,3,8(3 IF DATA FROM SCRAPS)
C   DATA=DATA MATRIX(10 PER WORD)
C   ITRASP=IF EQ.0,DATA ENTERED BY SPECIES
C   ITRASP=IF GT.0,DATA ENTERED BY RELEVES
C   ICOOC=CO-OCCURRENCE MATRIX
C   RID=REDUNDANCY MATRIX RID(I,K) IS THE REDUNDANCY
C       OF TABLE, RESULTING FROM MERGING TABLES I AND K
C   NSPEC=UP TO 150 SPECIES
C   NQUAD=UP TO 250 RELEVES
C   IFPR=IF GT.0,DETAILED PRINT OF THE CALCULATION IS PERFORMED
C   MTITLE=COMPUTATION HEADLINE
C   ITITLE=SINGLE TABLE HEADLINE
C   NT=NUMERICAL TABLE INDEX
C   NOUN=ALPHABETIC LABEL OF SPECIES
C   ISPEC=NUMERICAL LABEL OF SPECIES
C   NUDATA=NUMBER OF DATA IN RELEVES
C   NPAIR=IF GT.0,NPAIR PAIRS(IP,IIP) ARE INPUT
C   IP,IIP=NUMERICAL LABELS FOR TWO TABLES TO BE COMPARED.ONLY
C       RID(IP,IIP) COMPUTED
C   IENT=IF GT.0,ENTROPY OF RESULTING DATA TABLE IS COMPUTED
C   IFRE=IF GT.0,A TWO-COLUMN MATRIX OF FREQUENCIES FORMED.IENT SET
C       TO 1 AND CO-OCCURRENCE MATRIX NOT COMPUTED
C   INOTAB=IF GT.0,PRINTS TABLES OF THE STACK
C   NEWDIAG=IF GT.0, CO-OCCURRENCE MATRIX DIAGONAL ELEMENTS REPLACED BY
C           SPECIES FREQUENCY IN DATA TABLE
C   MLTAX3 WRITES ON TAPE4 RID MATRIX, BY ROWS(NO DIAGONAL ELEMENTS)
C     WITH FORMAT (16F5.3)
C
       INT=INFILE
       REWIND1
       REWIND 4
       REWIND 3
       RAND=RANF(BLA)
19     READ(5,1) (MTITLE(I),I=1,4),IFPR,IENT,NPAIR,IFRE,INOTAB
      1,NEWDIAG
       IF(MTITLE(1).EQ.10HFINEMLTAX3) GO TO 5
20     WRITE(6,21) (MTITLE(I),I=1,4),IFPR,IENT,NPAIR,IFRE,INOTAB
      1,NEWDIAG
1      FORMAT(4A10,6I5)
21     FORMAT(1H1,4A10/* MLTAX3-IPR IH NPAIR IFRE INOTAB NEWDIAG*/1X,6I5)
       IF(IFRE.NE.0) IENT=1
       NT=0
9      READ(INT,1) (ITITLE(I),I=1,4)
       IF(ITITLE(1).EQ.10HFINESAMPLE) GO TO 1000
```

```
      IF(ITITLE(1).EQ.10HSCRAPS-MLT) GO TO 1000
   11 NT=NT+1
      READ(INT,2) NSPEC,NQUAD,NUDATA,ITRASP
    2 FORMAT(16I5)
      WRITE(6,10) (ITITLE(I),I=1,4),NSPEC,NQUAD,NUDATA,ITRASP
10    FORMAT(1X,4A10/* MLTAX3-NSP NQ NDT ITR*/1X,6I5)
      DO 3 I=1,NSPEC
      READ(INT,4) ISPEC(I),ILCOPI(I),(NOUN(K),K=1,4)
4     FORMAT(I3,I7,4A10)
      IF(INOTAB.GT.0)WRITE(6,12)NT,I,ISPEC(I),ILCOPI(I),(NOUN(K),K=1,4)
3     CONTINUE
12    FORMAT(2X,4I8,2X,4A10)
      WRITE(1) NT,NSPEC,NQUAD,IFRE,(ILCOPI(I),I=1,NSPEC)
      CALL LFTR(NSPEC,NQUAD,NUDATA,NT,ITRASP,IFRE,INOTAB)
      GO TO 9
1000  WRITE(6,1001) ITITLE(1)
1001  FORMAT(1X////*  ALL TABLES LOADED-MERGING BEGINS.  MLTAX3*,A10)
      REWIND1
      REWIND 2
      DO 18 JJ=1,60
      DO 18 JK=1,60
18    RID(JJ,JK)=0.
      IF(NPAIR.EQ.0) GO TO 13
      DO 14 JJ=1,NPAIR
      READ(5,2) IP,IIP
      IF(IP.LE.IIP) GO TO 15
      IRKA=IIP
      IIP=IP
      IP=IRKA
15    CALL MELT(IP,IIP,NT,NUMAX,LMAX,ILCOPI)
      CALL STRUK(NUMAX,LMAX,NMED,IENT,IFPR,IP,IIP)
      IF(IFRE.GT.0) GO TO 14
      IF(IENT.GT.0) GO TO 14
      CALL CPCOOC(NUMAX,LMAX,NEWDIAG)
      CALL CPENTR(ICOOC,NUMAX,LMAX,TMARG,RID,IP,IIP,IFPR)
14    CONTINUE
      GO TO 16
   13 DO 17 JJ=1,NT
      DO 17 KK=JJ,NT
      CALL MELT(JJ,KK,NT,NUMAX,LMAX,ILCOPI)
      CALL STRUK(NUMAX,LMAX,NMED,IENT,IFPR,JJ,KK)
      IF(IFRE.GT.0) GO TO 17
      IF(IENT.GT.0) GO TO 17
      CALL CPCOOC(NUMAX,LMAX,NEWDIAG)
      CALL CPENTR(ICOOC,NUMAX,LMAX,TMARG,RID,JJ,KK,IFPR)
17    CONTINUE
```

```
16    WRITE(6,1002)
1002  FORMAT(1X////*   REDUNDANCY MATRIX-MLTAX3*)
      DO 6 K=2,NT
      WRITE(6,8) K
   8  FORMAT(* ROW=*,I4)
      KM1=K-1
      WRITE(4,27) (RID(K,I),I=1,KM1)
27    FORMAT(16F5.3)
6     WRITE(6,7)(RID(K,I),I=1,KM1)
7     FORMAT(10X,20F6.3)
      GO TO 19
5     REWIND3
      REWIND 2
      REWIND 1
      REWIND 4
      WRITE(6,100) MTITLE(1)
100   FORMAT(*0 WORK REGULARLY TERMINATED-MLTAX3*,A10)
      END
      SUBROUTINE LFTR(NSPEC,NQUAD,NUDATA,NTAB,ITRASP,IFRE,INOTAB)
      COMMON DATA(3750),ICOOC(11325),IBUF(250),ISPEC(150),ISOL(150),TMAR
     1G(150),RID(60,60),INT
      DIMENSION STR(250),IF(7)
C   IF=INPUT FORMAT FOR DATA TABLES
C   IND=THEIR INPUT FILE-IF NOT SPECIFIED,IND=INT
      READ(INT,100) IF,IND
100   FORMAT(7A10,5X,I5)
      IF(IND.LE.0) IND=INT
      PRINT 101,IF,IND
101   FORMAT(1X,7A10,I5)
      IF(ITRASP.GT.0) GO TO 1
      DO 2 I=1,NSPEC
      READ(IND,IF) (STR(N),N=1,NQUAD)
      NCOUNT=0
      DO 8 MAH=1,NQUAD
      IF(IFRE.LE.0) GO TO 8
      IF(STR(MAH).NE.0.) NCOUNT=NCOUNT+1
8     IBUF(MAH)=STR(MAH)
      IF(INOTAB.GT.0) WRITE(6,7) NTAB,(IBUF(N),N=1,NQUAD)
7     FORMAT(1X,40I3)
      LUNG=NQUAD
      IF(IFRE.LE.0) GO TO 2
      IBUF(1)=FLOAT(NCOUNT)*63./FLOAT(NQUAD)+0.5
      LUNG=1
2     WRITE(1) NTAB,(IBUF(N),N=1,LUNG)
      RETURN
   1  REWIND2
```

```
      DO 4 I=1,NQUAD
      READ(IND,IF) (STR(N),N=1,NUDATA)
      DO 3 NN=1,NSPEC
      MM=ISPEC(NN)
    3 IBUF(NN)=STR(MM)
    4 WRITE(2) (IBUF(N),N=1,NSPEC)
      DO 5 J=1,NSPEC
      REWIND2
      NCOUNT=0
      DO 6 I=1,NQUAD
      READ(2) (IBUF(L),L=1,NSPEC)
      IF(IFRE.LE.0) GO TO 6
      IF(IBUF(J).NE.0) NCOUNT=NCOUNT+1
    6 ICOOC(I)=IBUF(J)
      IF(INOTAB.GT.0) WRITE(6,7) NTAB,(ICDOC(L),L=1,NQUAD)
      LUNG=NQUAD
      IF(IFRE.LE.0) GO TO 5
      ICOOC(1)=FLOAT(NCOUNT)*63./FLOAT(NQUAD)+0.5
      LUNG=1
    5 WRITE(1) NTAB,(ICOOC(L),L=1,LUNG)
      RETURN
      END
      SUBROUTINE MELT(IR,IC,ITAB,NUMAX,LMAX,ILCOPI)
      COMMON DATA(3750),ICOOC(11325),IBUF(250),ISPEC(150),ISOL(150),TMAR
     1G(150),RID(60,60),INT
      DIMENSION NOME(150),LP(150),LPP(150),NRDM(250),ILCOPI(1)
      IF(IR.GT.ITAB.OR.IC.GT.ITAB) RETURN
      REWIND1
      REWIND 2
      I=1
        ILIB=1
        NUMAX=0
        IIFLAG=0
      DO 100 MM=1,150
      LP(MM)=0
      LPP(MM)=0
  100 NOME(MM)=0
      DO 200  MM=1,250
  200 NRDM(MM)=0
    1 READ(1) NUT,NSPEC,NQUAD,IFRE,(ILCOPI(II),II=1,NSPEC)
      IF(IFRE.NE.0) NQUAD=1
      IF(NUT.EQ.IR) GO TO 2
      IF(NUT.EQ.IC) GO TO 3
      DO 4 LI=1,NSPEC
    4 READ(1) NUT,(IBUF(L),L=1,NQUAD)
      I=I+1
```

```
      IF(I.LE.ITAB) GO TO 1
      RETURN
    2 DO 5 MI=1,NSPEC
      NOME(MI)=ILCOPI(MI)
      LP(MI)=ILIB
      READ(1) NUT,(IBUF(L),L=1,NQUAD)
      DO 6 J=1,NQUAD
      CALL MOVCHR(10,IBUF(J),ILIB,DATA)
      ILIB=ILIB+1
      IF(ILIB.LT.37500) GO TO 6
      WRITE(6,23)
   23 FORMAT(*  TOO MANY DATA IN FIRST TABLE-MLTAX3*)
      GO TO 7
    6 CONTINUE
    5 CONTINUE
      LUNI=NQUAD
      LUNII=LUNI
      LMAX=NQUAD
      NUMAX=NSPEC
      NOMAX=NUMAX
      IF(IC.NE.IR) GO TO 1
      GO TO 29
    3 DO 8 I=1,NSPEC
      LUNII=NQUAD
      DO 9 K=1,NOMAX
      IF(ILCOPI(I).EQ.NOME(K)) GO TO 10
    9 CONTINUE
      LMAX=MAX0(LMAX,NQUAD)
      LPP(NUMAX+1)=ILIB
      NOME(NUMAX+1)=ILCOPI(I)
      NUMAX=NUMAX+1
      IF(NUMAX.LE.150) GO TO 11
      WRITE(6,24)
   24 FORMAT(*  TOO MANY SPECIES-MLTAX3*)
      GO TO 29
   10 LPP(K)=ILIB
      IF(IIFLAG.GT.0) GO TO 11
      IIFLAG=2
   11 READ(1) NUT,(IBUF(LL),LL=1,NQUAD)
      DO 12 J=1,NQUAD
      CALL MOVCHR(10,IBUF(J),ILIB,DATA)
      ILIB=ILIB+1
      IF(ILIB.LT.37500) GO TO 12
      WRITE(6,25)
   25 FORMAT(*  TOO MANY DATA DUE TO SECOND TABLE-MLTAX3*)
      GO TO 29
```

```
   12 CONTINUE
    8 CONTINUE
29    ILPIUC=MINO(LUNI,LUNII)
      ILPIUL=MAXO(LUNI,LUNII)
      IF(IC.NE.IR) LMAX=2*ILPIUC
      IF(LUNI.EQ.LUNII) GO TO 7
      DO 13 IRDM=1,ILPIUC
28    NUPA=RANF(BURP)*FLOAT(ILPIUL)+0.5
      DO 27 IOI=1,IRDM
      IF(NUPA.EQ.NRDM(IOI)) GO TO 28
27    CONTINUE
      NRDM(IRDM)=NUPA
13    CONTINUE
      WRITE(6,15) ILPIUC
15    FORMAT(* *////////* NEW COUPLE-NARROWER TABLE WIDTH*,I4)
      WRITE(6,26) (NRDM(MN),MN=1,ILPIUC)
26    FORMAT(* RANDOM POINTERS IN WIDER TABLE*/1X,25I4)
    7 WRITE(6,14) IR,IC
14    FORMAT(*0TABLES  *,I3,*  AND  *,I3)
      DO 17 K=1,NUMAX
      DO 16 I=1,LMAX
   16 IBUF(I)=0
      IF(LP(K).EQ.0) GO TO 18
      IPUN=LP(K)
      DO 19 J=1,ILPIUC
      IF(LUNI.GT.LUNII) IPUN=LP(K)+NRDM(J)-1
      CALL MOVCHR(IPUN,DATA,10,IBUF(J))
   19 IPUN=IPUN+1
   18 IF(LPP(K).EQ.0) GO TO 20
      IPUN=LPP(K)
      JI=ILPIUC+1
      DO 21 J=1,ILPIUC
      IF(LUNII.GT.LUNI) IPUN=LPP(K)+NRDM(J)-1
      CALL MOVCHR(IPUN,DATA,10,IBUF(JI))
      JI=JI+1
   21 IPUN=IPUN+1
   20 WRITE(2) (IBUF(J),J=1,LMAX)
   17 CONTINUE
      RETURN
      END
      SUBROUTINE STRUK(NSPEC,NQUAD,NMED,IENT,IFPR,NI,NII)
      COMMON DATA(3750),ICOOC(11325),IBUF(250),ISPEC(150),ISOL(150),TMAR
     1G(150),RID(60,60),INT
      DIMENSION QMARG(250)
      REWIND2
      WRITE(6,800)
```

```
  800   FORMAT(1X////*   CALCULATIONS ON DATA MATRIX-MLTAX3*)
        TG=0.
        DO 4 JJ=1,NSPEC
        ISOL(JJ)=0
      4 TMARG(JJ)=0.
        DO 5 JJ=1,NQUAD
      5 QMARG(JJ)=0.
        IF(IFPR.GT.0) WRITE(6,400)
  400   FORMAT(1X////*   DATA TABLE-MLTAX3*)
        NMED=0
        K=1
        DO 1 I=1,NSPEC
        IF(IFPR.GT.0) WRITE(6,500) I
    500 FORMAT(* SPEC*,I5)
        READ(2) (IBUF(L),L=1,NQUAD)
        IF(IFPR.GT.0) WRITE(6,300) (IBUF(L),L=1,NQUAD)
    300 FORMAT(6X,10I3,3X,10I3,3X,10I3,3X,10I3)
        DO 10 JJ=1,NQUAD
        QMARG(JJ)=QMARG(JJ)+IBUF(JJ)
        TMARG(I)=TMARG(I)+IBUF(JJ)
        TG=TG+IBUF(JJ)
     10 CONTINUE
        DO 2 J=1,NQUAD
        CALL MOVCHR(10,IBUF(J),K,DATA)
        K=K+1
        IF(K.GT.37500) GO TO 3
      2 CONTINUE
      1 CONTINUE
        INDIC=0
          ISOSTA=0
        DO 600 IU=1,NQUAD
        ISUM=0
        DO 700 IA=1,NSPEC
        IO=(IA-1)*NQUAD+IU
        CALL MOVCHR(IO,DATA,10,ISOSTA)
        IF(ISOSTA.LE.0) GO TO 700
        IDSUM=1
        ISUM=ISUM+IDSUM
        INDIC=IA
  700   CONTINUE
        IF(INDIC.GT.0.AND.ISUM.EQ.1) ISOL(INDIC)=ISOL(INDIC)+1
        NMED=NMED+ISUM
  600   CONTINUE
        IF(IENT.LE.0) RETURN
        NTOT=NQUAD*NSPEC
        HXY=0.
```

```
      DO 6 JJ=1,NTOT
      IPIRK=0
      CALL MOVCHR(JJ,DATA,10,IPIRK)
      PIK=FLOAT(IPIRK)/TG
      IF(PIK.LE.0.) PIK=1.
      DHXY=PIK*ALOG(PIK)
6     HXY=HXY+DHXY
      HY=0.
      DO 7 JJ=1,NSPEC
      PIK=TMARG(JJ)/TG
      IF(PIK.LE.0.) PIK=1.
      IF(IFPR.GT.0) WRITE(6,13) JJ,TMARG(JJ)
13    FORMAT(* SPECIES N.*,I4,5X,*MARGINAL TOT.=*,F10.0)
      DHY=PIK*ALOG(PIK)
7     HY=HY+DHY
      HX=0.
      DO 8 JJ=1,NQUAD
      PIK=QMARG(JJ)/TG
      IF(PIK.LE.0.) PIK=1.
      IF(IFPR.GT.0) WRITE(6,14) JJ,QMARG(JJ)
14    FORMAT(* QUADRAT N.*,I4,5X,*MARGINAL TOT.=*,F10.0)
      DHX=PIK*ALOG(PIK)
8     HX=HX+DHX
      H=-(HX+HY-HXY)
      HMAX=ALOG(FLOAT(NSPEC))
      RIDOND=1.-H/HMAX
      RID(NI,NII)=RIDOND
      RID(NII,NI)=RIDOND
      HTAB=0.
       RITAB=0.
       ENM=NSPEC
      IF((NMED/NQUAD).EQ.NSPEC) GO TO 12
      ENM=NMED/NQUAD
      HTAB=ENM*ALOG(FLOAT(NSPEC)/ENM)
      RITAB=1.-H/HTAB
12    WRITE(6,9) HX,HY,HXY,H,HMAX,RIDOND,HTAB,RITAB,ENM
9     FORMAT(* HX=*,E10.3/* HY=*,E10.3/* HXY=*,E10.3/* H=*,E10.3/* HMAX=
     1*,E10.3/* 1-H/HMAX=*,E10.3/* N  LN (NSPEC/N)=*,E10.3/* 1-H/HTAB=*,
     2E10.3/* MEAN SPEC.NUMB./QUADR.=*,F10.0)
      RETURN
    3 WRITE(6,200) I,J
200   FORMAT(*  TOO MANY DATA IN MLTAX3-RELN.*,I5,*SPN.*,I5)
      RETURN
      END
      SUBROUTINE CPCOOC(NSPEC,NQUAD,ND)
```

124

```
      COMMON DATA(3750),ICOOC(11325),IBUF(250),ISPEC(150),ISOL(150),TMAR
     1G(150),RID(60,60),INT
      NPOST=NSPEC*(NSPEC+1)/2
      DO 3 I=1,NPOST
    3 ICOOC(I)=0
      DO 1 I=1,NSPEC
      DO 1 K=I,NSPEC
      IPRSC=0
      IND=ICOL(NSPEC,I)+K-(I-1)
      IF(ND.GT.0)GO TO 5
      IF(K.EQ.I) GO TO 4
    5 DO 2 J=1,NQUAD
      NIJ=(I-1)*NQUAD+J
      NJK=(K-1)*NQUAD+J
      IPARK=0
        IPORK=0
      CALL MOVCHR(NIJ,DATA,10,IPARK)
      CALL MOVCHR(NJK,DATA,10,IPORK)
      IF(IPARK.GT.0) IPARK=1
      IF(IPORK.GT.0) IPORK=1
      IDPRSC=IPARK*IPORK
    2 IPRSC=IPRSC+IDPRSC
      ICOOC(IND)=IPRSC
      GO TO 1
    4 IF(K.EQ.I) ICOOC(IND)=ISOL(I)
    1 CONTINUE
      RETURN
      END
      SUBROUTINE CPENTR(ICOOC,NSPEC,NQUAD,TMARG,RID,JK,KJ,IFPR)
      DIMENSION ICOOC(1),TMARG(1),RID(60,60)
      WRITE(6,11)
   11 FORMAT(1X////*  CALCULATIONS ON CO-OCCURRENCE MATRIX-MLTAX3*)
      DO 4 I=1,NSPEC
      TMARG(I)=0.
      TINCR=0
      DO 1 K=1,NSPEC
      IF(K.LT.I) GO TO 2
      IROW=I
      KOL=K
      GO TO 3
    2 IROW=K
      KOL=I
    3 IND=ICOL(NSPEC,IROW)+KOL-(IROW-1)
      TINCR=ICOOC(IND)
    1 TMARG(I)=TMARG(I)+TINCR
      IF(IFPR.GT.0) WRITE(6,5) I,TMARG(I)
```

```
4      CONTINUE
    5 FORMAT(1X,7HROW.N.=,I5,10X,10HMARG.TOT.=,F10.0)
      TOTG=0.
      D010 I=1,NSPEC
10     TOTG=TOTG+TMARG(I)
      IF(IFPR.GT.0) WRITE(6,6) TOTG
    6 FORMAT(1X, 9HGEN.TOT.=,F10.0////1X)
      HMARG=0.
      DO 7 I=1,NSPEC
      PI=TMARG(I)/TOTG
      IF(PI.LE.0.) PI=1.
      PILOG2=PI*ALOG(PI)
    7 HMARG=HMARG+PILOG2
      HMARG=HMARG*2.
      HINT=0.
      DO 8 I=1,NSPEC
      DO 8 K=I,NSPEC
      IND=ICOL(NSPEC,I)+K-(I-1)
      PIK=ICOOC(IND)/TOTG
      IF(PIK.LE.0.) PIK=1.
      PIKLOG2=PIK*ALOG(PIK)
      IF(K.NE.I) PIKLOG2=PIKLOG2*2.
    8 HINT=HINT+PIKLOG2
      ENTR=-(HMARG-HINT)
      HMAX=ALOG(FLOAT(NSPEC))
      RIDOND=1.-(ENTR/HMAX)
      WRITE(6,9) HMARG,HINT,ENTR,HMAX,RIDOND
9      FORMAT(* HX+HY=*,E10.3/* HXY=*,E10.3/* H=*,E10.3/* HMAX=*,E10.3/*
      11-H/HMAX=*,E10.3)
      RID(JK,KJ)=RIDOND
      RID(KJ,JK)=RID(JK,KJ)
      RETURN
      END
      FUNCTION ICOL(NUM,IRIG)
      ICOL=0
      IF(IRIG.EQ.1) RETURN
      LARIG=IRIG-1
      DO 1 MM=1,LARIG
    1 ICOL=ICOL+NUM-MM+1
      RETURN
      END
      OVERLAY(4,0)
      PROGRAM SCRAPS
C    TABLE PRODUCED IN MLTAX1 SUBDIVIDED INTO NSOTT SUBMATRICES FOR
C    INPUT IN MLTAX2 AND MLTAX3
C    IC,IIC=COLUMN SUBSCRIPTS OF FIRST(1,1) AND LAST ELEMENT(X,Y) IN A
```

```
C    SUBMATRIX OF DIMENSIONS(X*Y)
C      IR,IIR=CORRESPONDING ROW SUBSCRIPTS
C    LESPEC=ILCOPI IN MLTAX1
C    ARIG=NUMBER OF ROWS IN MATRIX PRODUCED BY MLTAX1
C    LIN=INPUT FILE FOR SCRAPS(5 OR 3)
C    LOUT=OUTPUT FILE FOR SCRAPS(7 OR 3)
C    LET=NUMBER OF TABLES SUPPLIED TO SCRAPS BY MLTAX1
     COMMON/COMUNE/INFILE    ,IFCUT,LIN,LOUT,LET
     DIMENSION IC(60),IIC(60),IR(60),IIR(60),LESPEC(118),LARIG(118),ILT
    1T(4),NALFA(4),ILFODA(8),TAFOLP(11800),ARIG(118)
     REWIND 1
     IF(LIN.NE.5) REWIND LIN
     IF(LOUT.NE.7) REWIND LOUT
     INTMD=1
     IF(LIN.NE.3.OR.LOUT.NE.3) INTMD=LOUT
     NALFA(1)=10HSCRAPS-MLT
       NALFA(2)=10HAX1
     NALFA(3)=10H
       NALFA(4)=10H
     ILFODA(1)=10H(40F3.0)
     DO 6 J=2,8
6    ILFODA(J)=10H
     DO 1 I=1,LET
     READ(5,2) NSOTT
     READ(5,2) (IR(J),IC(J),IIR(J),IIC(J),J=1,NSOTT)
2    FORMAT(16I5)
     DO 16 J=1,NSOTT
16   WRITE(6,17) IR(J),IC(J),IIR(J),IIC(J),J,I
17   FORMAT(* SCRAPS-CORNERS*,4I5,* SBTAB *,I5,* OF TABLE *,I5)
     READ(LIN,100) NUT,NS,NQ,(ILTT(J),J=1,4)
100  FORMAT(2X,3I8,2X,4A10)
     READ(LIN,200) (LESPEC(J),J=1,NS)
200  FORMAT(10I8)
     LL=1
     DO 3 L=1,NQ
     READ(LIN,600) (LARIG(J),J=1,NS)
600  FORMAT(40I2)
     DO 4 J=1,NS
     CALL MOVCHR(10,LARIG(J),LL,TAFOLP)
     LL=LL+1
4    CONTINUE
3    CONTINUE
     DO 5 M=1,NSOTT
     NSP=IIC(M)-IC(M)+1
     NQD=IIR(M)-IR(M)+1
     NDT=NSP
```

```
      ITRP=1
      IPR=0
        IHH=1
      WRITE(INTMD,300) (ILTT(J),J=1,4)
300   FORMAT(4A10)
      WRITE(INTMD,2) NSP,NQD,NDT,ITRP,IPR,IHH
      L1=IC(M)
        L2=IIC(M)
        L3=IR(M)
        L4=IIR(M)
      DO 15 N=L1,L2
      NIC=N-L1+1
15    WRITE(INTMD,400) NIC,LESPEC(N),(NALFA(J),J=1,4)
400   FORMAT(I3,I7,4A10)
      WRITE(INTMD,500) (ILFODA(J),J=1,8)
500   FORMAT(8A10)
      DO 7 K=L3,L4
      DO 8 J=L1,L2
      KARPO=(K-1)*NS+J
      CALL MOVCHR(KARPO,TAFOLP,10,LARIG(J-L1+1))
8     ARIG(J-L1+1)=LARIG(J-L1+1)
7     WRITE(INTMD,601) ( ARIG(JJ),JJ=1,NSP)
601   FORMAT(40F3.0)
5     CONTINUE
1     CONTINUE
C   IT HAS WRITTEN ALL SCRAPS OF ALL MLTAX1-TABLES ON INTERMEDIATE FILE-
C   IT IS 1,IF LIN=3 AND LOUT=3,OR EQ LOUT IF OTHERWISE
      WRITE(INTMD,300) (NALFA(I),I=1,4)
      IF(LIN.NE.5) REWIND LIN
      IF(INTMD.NE.7) REWIND INTMD
      IF(LIN.NE.3.OR.LOUT.NE.3) GO TO 9
C   IT TRANSFERS OUTPUT DATA FROM FILE INTMD=1 TO FILE
C  LOUT=3 - FILES HAVE ALREADY BEEN REWINDED
13    READ(INTMD,300) (ILTT(J),J=1,4)
      IF(ILTT(1).EQ.NALFA(1)) GO TO 14
12     WRITE(LOUT,300) (ILTT(J),J=1,4)
      READ(INTMD,2) NSP,NQD,NDT,ITRP,IPR,IHH
      WRITE(LOUT,2) NSP,NQD,NDT,ITRP,IPR,IHH
      DO 10 J=1,NSP
      READ(INTMD,400) NIC,LESPEC(J),(NALFA(JJ),JJ=1,4)
10    WRITE(LOUT,400) NIC,LESPEC(J),(NALFA(JJ),JJ=1,4)
      READ(INTMD,500) (ILFODA(JJ),JJ=1,8)
      WRITE(LOUT,500) (ILFODA(JJ),JJ=1,8)
      DO 11 J=1,NQD
      READ(INTMD,601) ( ARIG(JJ),JJ=1,NSP)
11    WRITE(LOUT,601) ( ARIG(JJ),JJ=1,NSP)
```

```
         GO TO 13
14       WRITE(LOUT,300) (NALFA(I),I=1,4)
         REWIND  LOUT
             REWIND INTMD
9        READ(5,300) LAFINE
         WRITE(6,700) LAFINE
700      FORMAT(*OSCRAPS HAS REGULARLY TERMINATED*,A10)
         END
CCINF1F
         PROGRAM CINF1F(INPUT=100,OUTPUT=100,TAPE5=INPUT,TAPE1=514,TAPE2=51
        14)
         INTEGER H(100,2),B,G1,A(100),CDOL,G2,D,F,X(100,100),C(100),V
         DIMENSION          Y(100),R(100,100),E(100,100),   G(100)
        2,IFMT(8),LFMT(8)



C     M=NUMBER OF ROWS(UP TO 100)
C     N=NUMBER OF COLUMNS(UP TO 100)
C     CDOL=OPTION FOR CONTINUOUS VARIABLES
C          IF=CONT,CONTINUOUS VARIABLES ARE DISCRETIZED
C     T1=CLASS INTERVAL IN STANDARD DEVIATION UNITS
C     INF=INPUT FILE(CARDS OR TAPE1 OR TAPE2)
C     IFBLA=OUTPUT FILE FOR SEQUENCE OF CLASSIFIED OBJECTS
C     ITR=IF GT.0,DATA ENTERED BY RELEVES
C     IFMT=INPUT FORMAT FOR DATA TABLE(FORMAT (8A10))
C     LFMT=FORMAT FOR PRINTING DATA TABLES(FORMAT (8A10))
C     X=DATA MATRIX(INTEGER NUMBERS)
C     +++++++++++++++++++++++++++++++++++++++++++++++++++++++++++++
      PRINT 9000
9000  FORMAT (1HT)
1111  READ 1,M,N,CDOL,T1,INF,IFBLA,ITR
1     FORMAT(2I5,A5,F5.0,12I5)
      IF (EOF(5)) 1981,8
8     READ 3,IFMT
      READ 3,LFMT
3     FORMAT (8A10)
      PRINT 4444 ,IFMT,LFMT
4444  FORMAT (//* IFMT = *,8A10/* LFMT = *,8A10)
      PRINT 2,M,N
2     FORMAT (1H1,20X,*PROGRAM CINF1F*//
     2*.REFERENCE: INF1, ORLOCI, L. 1969. NATURE, VOL. 223*/
     3* NUMBER OF ROWS IN DATA MATRIX =    M =*,I5/
     4* NUMBER OF COLUMNS IN DATA MATRIX = N =*,I5)
      PRINT 70 ,CDOL
70    FORMAT (* CDOL= *,A4)
      IF (CDOL.NE.4HCONT) GOTO 130
```

```
      PRINT 100 , T1
100   FORMAT (*  CLASS INTERVAL IN STANDARD*/
     2* DEVIATION UNITS=*,F5.0)
130   IF(ITR.GT.0)GO TO 2002
      READ(INF,IFMT)((X(I,J),J=1,N),I=1,M)
      GO TO 2003
2002  READ(INF,IFMT)((X(I,J),I=1,M),J=1,N)
2003  CONTINUE
      PRINT LFMT, ((X(I,J),J=1,N ),I=1,M)
      DO 240 IY=1,100
240   Y(IY)=0
      DO 250 IR=1,N
      DO 250 JR=1,N
250   R(IR,JR)=0
      DO 300 K=1,N
      H(K,1)=K
      H(K,2)=0
300   R(K,1)=K
      K=N
      D=1
C
C--UPDATE C ARRAY

330   KI=K-1
      DO 9001 IUK=1,N
      DO 9001 IAK=1,N
9001  E(IUK,IAK)=0
      DO 1000 IH=1,KI
      IHH=IH+1
      DO 1000 I1=IHH,K
      W=0
      IE=0
      DO 450 J=1,N
      M1=R(IH,J)
      IF (M1.EQ.0) GOTO 460
      IE=IE+1
450   C(IE)=M1
460   DO 510 J=1,N
      M1=R(I1,J)
      IF (M1.EQ.0) GOTO 560
      IE=IE+1
 510  C(IE)=M1
C
C--COMPUTE MEAN AND STANDARD DEVIATION FOR ROWS WITHIN SUBSETS
C--SORT OBSERVATIONS INTO FREQUENCY CLASSES WITHIN ROWS
C--CLASS INTERVAL IS SET AT 1/4 OF STANDARD DEVIATION
```

130

```
C--COMPUTE INFORMATION MATRIX

  560   DO 960 B=1,M
        DO 600 J=1,IE
        M1=C(J)
  600   G(J)=X(B,M1)
        G1=IE
        GG=Z=0
        IF (CDOL.EQ.4HCONT) GOTO 670
        DO 650 IAG=1,100
  650   A(IAG)=G(IAG)
        GOTO 790
  670   DO 700 L=1,G1
        GG=GG+G(L)
  700   Z=Z+G(L)**2
        Z=Z-GG**2/G1
        GG=GG/G1
        DO 730 IA=1,100
  730   A(IA)=0
        IF (Z.LE.0) GOTO 790
        Z=Z/(G1-1)
        DO 780 L=1,G1
  780   A(L)=INT(G(L)/(T1*SQRT(Z)))
  790   V=-1000
        Z1=0
        G2=G1-1
        DO 930 L=1,G2
        I=A(L)
        K1=1
        IF (I.EQ.V) GOTO 930
        LL=L+1
        DO 910 J=LL,G1
        F=A(J)
        IF (F.EQ.V) GOTO 910
        IF (F.NE.I) GOTO 910
        K1=K1+1
        A(J)=V
  910   CONTINUE
        AK=K1
         AG=G1
        Z1=Z1-K1*ALOG(AK/AG)
  930   CONTINUE
        IF (A(G1).EQ.V) GOTO 960
        ONE=1
         AG=G1
        Z1=Z1-ALOG(ONE/AG)
```

```
 960   W=W+Z1
1000   E(IH,I1)=E(I1,IH)=W
       PRINT 1020 , D
1020   FORMAT (//* CLUSTERING PASS*,I3/)

C--FIND VALID UNIONS, UPDATE REGISTERS

       DO 1270 L=1,K
       S=T=1000000
       DO 1140 M1=1,K
       IF (L.EQ.M1) GOTO 1140
       W=E(L,M1)-Y(L)-Y(M1)
       IF (W.GE.S) GOTO 1140
       S=W
       I=M1
1140   CONTINUE
       DO 1210 M1=1,K
       IF (M1.EQ.I) GOTO 1210
       W=E(I,M1)-Y(I)-Y(M1)
       IF (W.GE.T) GOTO 1210
       T=W
       B=M1
1210   CONTINUE
       H(L,1)=L
       H(L,2)=0
       IF (B.NE.L) GOTO 1270
       H(L,2)=I
       Y(L)=E(L,I)
1270   CONTINUE

       DO 1410 L=1,K
       IF (H(L,2).LE.H(L,1)) GOTO 1410
       F=H(L,2)
       DO 1340 M1=1,N
       IF (R(L,M1).EQ.0) GOTO 1350
1340   IC=M1
1350   DO 1400 M1=1,N
       IF (R(F,M1).EQ.0) GOTO 1410
       IC=IC+1
       R(L,IC)=R(F,M1)
1400   R(F,M1)=0
```

```
1410  CONTINUE
      IC=0
      DO 1530 F=1,K
      IF (R(F,1).EQ.0) GOTO 1530
      IC=IC+1
      Y(IC)=Y(F)
      DO 1520 I=1,N
      IF (R(F,I).EQ.0) GOTO 1530
      R(IC,I)=R(F,I)
      IF (IC.EQ.F) GOTO 1520
      R(F,I)=0
1520  CONTINUE
1530  CONTINUE
      K=IC
      DO 1650 L=1,K
      NUM=0
      DO 1620 M1=1,N
      IF (R(L,M1).EQ.0)GOTO 1630
1620  NUM=NUM+1
1630  PRINT 2000 , L, (R(L,M1),M1=1,NUM)
1650  PRINT 3000 , Y(L)
2000  FORMAT(3X,*SUBSET*,I3,*COLUMN(S)--*,(3X,25F4.0))
3000  FORMAT (3X,*POOLED INFORMATION--*,G11.4)
      D=D+1
      IF(K.GT.1) GOTO 330
      PRINT 2004,IFBLA
2004  FORMAT(*0FINAL SEQUENCE ON FILE*,I4,* WITH FORMAT(5X,25I4)*)
      IF(IFBLA.GT.0)WRITE(IFBLA,2001)(R(K,M1),M1=1,N)
2001  FORMAT(5X,25F4.0)
      GOTO 1111
 1981 STOP
      END


CCINF2
      PROGRAM CINF2(INPUT=100,OUTPUT=100,TAPE5=INPUT,TAPE1=514,TAPE2=514
     1)
      INTEGER F(100,2),C(100)
      DIMENSION X(100,100),E(100,100),Y(100),R(100,100),BETA(100),G(100)
     1,AN(100),A(100),B(100),IFMT(8),LFMT(8),STAMPA(100)
      REAL N1
      PRINT 9000
9000  FORMAT (1HT)



C     M=NUMBER OF ROWS(UP TO 100)
C     N=NUMBER OF COLUMNS(UP TO 100)
```

```
C      INF=INPUT FILE FOR DATA TABLES(CARDS OR TAPE1 OR TAPE2)
C      IFBLA=OUTPUT FILE FOR SEQUENCE OF CLASSIFIED OBJECTS(TAPE1 OR TAPE
C      ITR=IF GT.0,DATA ENTERED BY RELEVES
C      IFMT=INPUT FORMAT FOR DATA TABLES
C      LFMT=LIST FORMAT FOR DATA TABLES
C      X=DATA MATRIX
C      ++++++++++++++++++++++++++++++++++++++++++++++++++++++++++++++++++
1955  READ 1,M,N,INF,IFBLA,ITR
      IF (EOF(5)) 1981,7
7     PRINT 2,M,N
1     FORMAT (16I5)
2     FORMAT (1H1,20X,*PROGRAM CINF2*//
     2* REFERENCE: INF2, ORLOCI, L. 1969, NATURE, VOL. 223*/
     3* NUMBER OF ROWS IN DATA MATRIX =    M =*,I5/
     4* NUMBER OF COLUMNS IN DATA MATRIX = N =*,I5/)
      READ 3, IFMT
      READ 3, LFMT
3     FORMAT (8A10)
      IF(ITR.GT.0)GO TO 2002
      DO 110 I=1,M
110   READ(INF,IFMT)(X(I,J),J=1,N)
      GO TO 2003
2002  READ(INF,IFMT)((X(I,J),I=1,M),J=1,N)
2003  CONTINUE

C--FIRST SUBSCRIPT OF X = LABEL FOR ROWS
C--SECOND SUBSCRIPT OF X = LABEL FOR COLUMNS

      DO 5 I=1,M
5     PRINT LFMT,(X(I,J),J=1,N)
      DO 250 I=1,100
      Y(I)=0
      DO 250 J=1,2
250   F(I,J)=0
      DO 260 I=1,N
      DO 260 J=1,N
260   R(I,J)=0
C-----------------------------------------------------------------
C--FIND COLUMN TOTALS (MAT AN), COUNT NON-ZERO VALUES
C--IN COLUMNS AND COMPUTE CONTRIBUTIONS (MAT A) OF COLUMNS
C--TO MUTUAL INFORMATION
C-----------------------------------------------------------------
      DO 450 L=1,N
      F(L,1)=L
      R(L,1)=L
      S=0
```

```
        IC=0
        W=0
        DO 410 M1=1,M
            IF(X(M1,L).EQ.0) GOTO 410
        IC=IC+1
        S=S+X(M1,L)
        W=W+X(M1,L)*ALOG(X(M1,L))
410     CONTINUE.
        AN(L)=S
        C(L)=IC
450     A(L)=W
        PRINT 460 ,(AN(L),L=1,N)
460     FORMAT(/* COLUMN TOTALS*//1X,(25F5.0))
        DO 530 L=1,N
        B(L)=G(L)=AN(L)
530     AN(L)=AN(L)*ALOG(AN(L))
        PRINT 560
560     FORMAT (//* NUMBER OF NON-ZERO VALUES IN COLUMNS*/)
        PRINT 590,(C(L),L=1,N)
590     FORMAT (1X,35I3 )
        G1=1
        K=N
C--GENERATE MUTUAL INFORMATION MATRIX

640     KN=K-1
        DO 820 NF=1,KN
        II=NF+1
        DO 820 IE=II,K
        W=A(NF)+A(IE)
        Z=AN(NF)+AN(IE)
            IF (G(NF)+G(IE).EQ.0) GOTO 730
        S=(G(NF)+G(IE))*ALOG(G(NF)+G(IE))
730     AG=0
        DO 780 L=1,M
        N1=X(L,NF)+X(L,IE)
            IF (N1.EQ.0) GOTO 780
        AG=AG+N1*ALOG(N1)
780     CONTINUE
        E(NF,IE)=ABS(W+S-Z-AG)
820     E(IE,NF)=E(NF,IE)


C--FIND VALID UNIONS, UPDATE REGISTERS

        DO 1060 L=1,K
        S=T=1000000
        DO 930 M1=1,K
```

```
          IF (L.EQ.M1) GOTO 930
      W=E(L,M1)-Y(L)-Y(M1)
          IF (W.GT.S) GOTO 930
      S=W
      I=M1
930   CONTINUE
      DO 1000 M1=1,K
          IF (M1.EQ.I) GOTO 1000
      W=E(I,M1)-Y(I)-Y(M1)
          IF (W.GT.T) GOTO 1000
      T=W
      IB=M1
1000  CONTINUE
      F(L,1)=L
      F(L,2)=0
          IF(IB.NE.L) GOTO 1060
      F(L,2)=I
      Y(L)=E(L,I)
1060  CONTINUE
      I=0
      DO 1390 L=1,K
          IF (F(L,2).LT.F(L,1)) GOTO 1280
      I=I+1
      NF=F(L,2)
      Y(I)=Y(L)
      F(I,1)=F(L,1)
      F(I,2)=F(L,2)
      DO 1210 M1=1,N
          IF (R(L,M1).EQ.0) GOTO 1220
      IC=M1
      R(I,M1)=R(L,M1)
          IF (I.EQ.L) GOTO 1210
      R(L,M1)=0
1210  CONTINUE
1220  DO 1270 M1=1,N
          IF (R(NF,M1).EQ.0) GOTO 1390
      IC=IC+1
      R(I,IC)=R(NF,M1)
1270  R(NF,M1)=0
1280      IF (F(L,2).GT.0) GOTO 1390
      I=I+1
      DO 1350 M1=1,N
          IF (R(L,M1).EQ.0) GOTO 1360
      R(I,M1)=R(L,M1)
          IF (I.EQ.L) GOTO 1350
      R(L,M1)=0
```

```
1350   CONTINUE
1360   F(I,1)=F(L,1)
       F(I,2)=F(L,2)
       Y(I)=Y(L)
1390   CONTINUE
       K=I
       DO 1620 L=1,K
       IC=0
       NF=F(L,1)
       I=F(L,2)
           IF (I.EQ.0) GOTO 1500
       AN(L)=AN(NF)+AN(I)
       A(L)=A(NF)+A(I)
       G(L)=G(NF)+G(I)
       GOTO 1530
1500   AN(L)=AN(NF)
       A(L)=A(NF)
       G(L)=G(NF)
1530   DO 1600 M1=1,M
           IF (I.EQ.0) GOTO 1570
       X(M1,L)=X(M1,NF)+X(M1,I)
       GOTO 1580
1570   X(M1,L)=X(M1,NF)
1580       IF (X(M1,L).EQ.0) GOTO 1600
       IC=IC+1
1600   CONTINUE
1620   C(L)=IC
       PRINT 1650,G1
1650   FORMAT (//* CLUSTERING PASS *,F4.0)
       DO 1920 L=1,K
       DO 33 NNN=1,25
33     STAMPA (NNN)=0
       NNN=0
       DO 1730 M1=1,N
           IF (R(L,M1).EQ.0) GOTO 1740
       NNN=NNN+1
1730   STAMPA(NNN)=R(L,M1)
1740   PRINT 2000,L,(STAMPA(KO),KO=1,NNN)
2000   FORMAT(*0 ' SUBSET*,I3,* : COLUMN(S)--*,(5X,25F4.0))
       DO 2222 NN=1,N
2222   BETA(NN)=0
       NUM=0
       DO 1800 M1=1,N
           IF (R(L,M1).EQ.0) GOTO 1810
       NUM=NUM+1
       IA=R(L,M1)+0.5
```

```
1800   BETA(M1)=B(IA)
1810   CONTINUE
       PRINT 3000,(BETA(IM),IM=1,NUM)
       PRINT 1820,(X(MM,L),MM=1,M)
       YY=2*Y(L)
1920   PRINT 1870,G(L),C(L),YY
3000   FORMAT (*   COLUMN TOTAL(S)--*,(20F5.0))
1820   FORMAT (*   ROW TOTALS--*,(20F5.0))
1870   FORMAT (*   GRAND TOTAL--*,F5.0,
      2* ,  NUMBER OF NON-ZERO VALUES--*, I5 ,* ,  TWICE MUTUAL*/
      3*   INFORMATION--*,G15.5)
       G1=G1+1
          IF (K.GT.1) GOTO 640
       PRINT 2004,IFBLA
2004   FORMAT(*0FINAL SEQUENCE ON FILE*,I4,* WITH FORMAT(5X,25F4.0)*)
       IF(IFBLA.GT.0)WRITE(IFBLA,2001)(STAMPA(KOK),KOK=1,NNN)
2001   FORMAT(5X,25F4.0)
       GOTO 1955
1981   STOP
       END


CCDAB
       PROGRAM CDAB(INPUT=100,OUTPUT=100,TAPE5=INPUT,TAPE1=514,TAPE2=514)
       DIMENSION IFMT(8),LFMT(8)
       REAL Y(100),E(100,100),G(100),IG
       INTEGER X(100,100),H(100,2),R(100,100),C(100),D,BUF(100)
      1,B,G1,Z,F
       PRINT 10000
10000  FORMAT (1HT)



C      M=NUMBER OF ROWS(UP TO 100)
C      N=NUMBER OF COLUMNS(UP TO 100)
C      INF=INPUT FILE FOR DATA TABLES(CARDS OR TAPE1 OR TAPE2)
C      IFBLA=OUTPUT FILE FOR FINAL SEQUENCE OF THE CLASSIFIED OBJECTS
C          (FORMAT (5X,25I4))
C      ITR=TRANSPOSITION OPTION(IF GT.0,DATA ENTERED BY RELEVES)
C      IFMT=INPUT FORMAT FOR DATA TABLES
C      LFMT=FORMAT FOR PRINTING DATA TABLES
C      X=DATA TABLE(INTEGER NUMBERS)
C      ............................................................
40     READ 10,M,N,INF,IFBLA,ITR
       IF(EOF(5))42,43
43     PRINT 20,M,N
       READ 50,IFMT
       READ 50,LFMT
```

```
50     FORMAT (8A10)
10     FORMAT (16I5)
20     FORMAT (1H1,*PROGRAM CDAB*//
      2* REFERENCE: INFA1E, ORLOCI, L. 1970, CAN J, BOT, 48*/
      3* NUMBER OF ROWS IN DATA MATRIX =    M =*,I5/
      4* NUMBER OF COLUMNS IN DATA MATRIX = N =*,I5)
       IF(ITR,GT,0)GO TO 2002
       DO 154 I=1,M
154    READ(INF,IFMT)(X(I,J),J=1,N)
       GO TO 2003
2002   DO 2004 J=1,N
2004   READ(INF,IFMT)(X(I,J),I=1,M)
2003   CONTINUE
C
       DO 1 I=1,M
1      PRINT LFMT, (X(I,J),J=1,N)
C
       DO 160 IND=1,100
       Y(IND)=0
       DO 160 IN=1,2
160    H(IND,IN)=0
       DO 170 IND=1,N
       DO 170 IN=1,N
170    R(IND,IN)=0
       DO 220  K=1,N
220    H(K,1)=R(K,1)=K
       K=N
       D=1
250    IK=K-1
       DO 600 IH=1,IK
         LH=IH+1
         DO 600 I1=LH,K
         W=0
         IE=0
         DO 340 J=1,N
         M1=R(IH,J)
           IF (M1,EQ,0) GOTO 350
           IE=IE+1
340      C(IE)=M1

350      DO 400 J=1,N
         M1=R(I1,J)
           IF (M1,EQ,0) GOTO 410
         IE=IE+1
400      C(IE)=M1
```

```
410      DO 570 B=1,M
           DO 450 J=1,IE
           M1=C(J)
450        G(J)=X(B,M1)
         G1=IE
         IG=0
         Z=0
           DO 510  L=1,G1
510        IG=IG+G(L)
         IG=IG/G1
           DO 570 L=1,G1
             IF (G(L).EQ.0) GOTO 570
           W=W+2*G(L)*ALOG(G(L)/IG)
570      CONTINUE
600      E(IH,I1)=E(I1,IH)=W
         PRINT 30, D
30       FORMAT (/* CLUSTERING PASS *,I3)
         DO 850 L=1,K
         S=1000000.0
         T=1000000.0
           DO 720 M1=1,K
             IF (L.EQ.M1) GOTO 720
           W=E(L,M1)-Y(L)-Y(M1)
             IF (W.GE.S) GOTO 720
           S=W
           I=M1
720      CONTINUE
         DO 790 M1=1,K
         IF (M1.EQ.I) GOTO 790
         W=E(I,M1)-Y(I)-Y(M1)
         IF (W.GE.T) GOTO 790
           T=W
           B=M1
790      CONTINUE
         H(L,1)=L
         H(L,2)=0
         IF (B.NE.L) GOTO 850
         H(L,2)=I
         Y(L)=E(L,I)
850      CONTINUE
         DO 990 L=1,K
         IF (H(L,2).LE.H(L,1)) GOTO 990
         F=H(L,2)
           DO 920 M1=1,N
           IF (R(L,M1).EQ.0) GOTO 930
920      IC=M1
```

140

```
930      DO 980 M1=1,N
         IF (R(F,M1).EQ.0) GOTO 990
         IC=IC+1
         R(L,IC)=R(F,M1)
980      R(F,M1)=0
990    CONTINUE
       IC=0
       DO 1110 F=1,K
       IF (R(F,1).EQ.0) GOTO 1110
       IC =IC+1
       Y(IC)=Y(F)
         DO 1100 I=1,N
         IF (R(F,I).EQ.0) GOTO 1110
         R(IC,I)=R(F,I)
         IF (IC.EQ.F) GOTO 1100
         R(F,I)=0
1100   CONTINUE
1110   CONTINUE
       K=IC
       DO 1200 L=1,K
       DO 3 NN=1,N
3      BUF(NN)=0
       NUM=0
       DO 1180 M1=1,N
       IF (R(L,M1).EQ.0) GOTO 1222
       BUF(M1)=R(L,M1)
1180   NUM=NUM+1
1222   PRINT 41,L,(BUF(M1),M1=1,NUM)
41     FORMAT(* --SUBSET *,I3/(8X,25I4))
       PRINT 2006,Y(L)
2006   FORMAT(1H+,110X,G10.3)
1200   CONTINUE
       D=D+1
       IF (K.GT.1) GOTO 250
       PRINT 2005,IFBLA
2005   FORMAT(*0FINAL SEQUENCE ON FILE *,I4/* WITH FORMAT(5X,25I4)*)
       IF(IFBLA.GT.0)  WRITE(IFBLA,2001) (BUF(NNN),NNN=1,N)
2001   FORMAT (5X,25I4)
       GOTO 40
42     STOP
       END
```

8. References

Anderberg, M.R. 1973. Cluster Analysis for Applications. Academic Press, London.

Basharin, G.P. 1959. On a statistical estimate for the entropy of a sequence of independent random variables. Theory Prob. Applic. 4: 333–336.

Bock, H.H. 1973. Automatische Classification. Van den Hoeck et Rpurecht, Göttingen.

Bottomley, J. 1971. Some statistical problems arising from the use of the information statistic in numerical classification. J. Ecol. 59: 339–342.

Clifford, H.T. & W. Stephenson. 1975. An Introduction to Numerical Classification. Academic Press, London.

Cormack, R.M. 1971. A review of classification. J. Roy. Statis. Soc. Series A 134: 324–353.

Fadeev, D.K. 1957. Zum Begriff der Entropie einer endlichen Wahrescheinlichkeitsschemas, Arbeiten für Informations theorie. I. Deutscher Verlag der Wissenschaften, Berlin.

Feinstein, A. 1958. Foundations of Information Theory. McGraw-Hill, New York.

Feoli, E. 1976. Correlation between single ecological variables and vegetation by means of cluster analysis. Not. Fitosoc. 12: 77–82.

Feoli, E. & L. Feoli Chiapella. 1979. Ranking relevés based on a sum of squares criterion. Vegetatio 39: 123–125.

Feoli, E., L. Feoli Chiapella, P. Ganis & A. Sorge. 1980. Spatial pattern analysis of abandoned grasslands in the Karst region of Trieste and Gorizia. Studia Geob. 1: 213–221.

Feoli, E. & M. Lagonegro. 1979. Intersection analysis in phytosociology: computer program and application. Vegetatio 40: 55–59.

Feoli, E. & D. Lausi. 1980. Hierarchical levels in syntaxonomy based on information functions. Vegetatio 42: 113–115.

Feoli, E. & M. Lagonegro. 1982. Syntaxonomical analysis of beech woods in the Apennines (Italy) using the program package IAHOPA. Vegetatio 50: 129–173.

Feoli, E., G. Parente & T. Trinco. 1979. I prati falciabili delle Valli del Natisone. I. Ordinamento e classificazione della vegetazione. Centro Regionale per la Sperimentazione Agraria, Udine, Italy.

Fisher, R.H. 1948. Statistical Methods for Research Workers. 10th edn. Oliver & Boyd, Edinburgh.

Fisher, R.A. 1963. Statistical Methods for Research Workers. 13th edn. Oliver & Boyd, Edinburgh.

Gallucci, V.F. 1973. On the principles of thermodynamics in ecology. Ann. Rev. Ecol. Syst. 4: 329–357.

Grassle, J.F., G.P. Patil, W. Smith & C. Tallie (eds.) 1979. Ecological diversity in theory and practice. Statistical Ecological Series, Vol. 6. International Co-operative Publishing House, Fairland, Maryland, U.S.A.

Greig-Smith, P. 1952. The use of random and contiguous quadrats in the study of the structure of plant communities. Ann. Bot. Lond. 16: 293–316.

Greig-Smith, P. 1964. Quantitative Plant Ecology. (2nd edn.). Butterworth, London.

Guillerm, J.L. 1971. Calcul de l'information fournie par un profil ecologique et valeur indicatrice des especes. Oecol. Plant 6: 209-225.

Hanski, I. 1978. Some comments on the measurement of niche metrics. Ecology 59: 168–174.

Hartigan, J.A. 1975. Clustering algorithms. Wiley, New York.

Hill, M.O. 1973. Diversity and evenness: a unifying notation and its consequences. Ecology 54: 427–432.

Hurd, L.E., M.V. Mellinger, L.L. Wolf & S.J. McNaughton. 1971. Stability and diversity at three trophic levels in terrestial successional ecosystems. Science 173: 1134–1136.

Hurlbert, S.H. 1971. The non-concept of species diversity: a critique and alternative parameters. Ecology 52: 577–586.

Jardine, N. & R. Sibson. 1971. Mathematical Taxonomy. Wiley, London.

Juhász-Nagy, P. 1976. Spatial dependence of plant populations. Part I. Equivalence analysis: an outline for a new model. Acta Bot. Acad. Sci. Hung. 22: 61–78.

Khinchin, A.I. 1957. Mathematical Foundations of Information Theory. Dover Publications, New York.

Kullback, S. 1959. Information Theory and Statistics. Wiley, New York.

Kushlan, J.A. 1976. Environmental stability and fish community diversity. Ecology 57: 821–825.

Lance, G.H. & W.T. Williams. 1968. Mixed data classificatory programs. II. Divisive systems. Aust. Comp. J. 1: 82–85.

Lagonegro, M. & E. Feoli. 1979. IAHOPA: a useful overlay-program for Systematics and Ecology. Quaderni del Centro di Calcolo 12, Università di Trieste.

Lagonegro, M. & E. Feoli. 1981. SINFUN: a program for information analysis. Quaderni del Centro di Calcolo 16, Università di Trieste.

Lausi, D. 1972. Die Logik der Pflanzensoziologischen Vegetationsanalyse – Ein Deutungsversuch. In: van der Maarel, E. & R. Tüxen (eds.), Basic Problems and Methods in Phytosociology, pp. 17–28. Junk, The Hague.

Maarel, E. van der. 1979. Multivariate methods in phytosociology with reference to the Netherlands. In: M.J.A. Werger (ed.), The Study of Vegetation, pp. 163–225. Junk, The Hague.

MacArthur, R.H. 1965. Patterns of species diversity. Biol. Rev. 40: 510–533.

Margalef, D.R. 1958. Information theory in Ecology. In: L. van Bertalanffy & A Rapoport (eds.), General Systems, Yearbook of the Society for General Systems Research 3: 36–71.

Orlóci, L. 1969a. Information theory models for hierarchic and non-hierarchic classifications. In: A.J. Cole (ed.), Numerical Taxonomy, pp. 148–164. Academic Press, London.

Orlóci, L. 1969b. Information analysis of structure in biological collections. Nature 223: 483–484.

Orlóci, L. 1970a. Automatic classification of plants based on the use of information content. Can. J. Bot. 48: 793–802.

Orlóci, L. 1970b. Analysis of vegetation samples based on the use of information. J. Theor. Biol. 29: 173–189.

Orlóci, L. 1971. An information theory model for pattern analysis. J. Ecol. 59: 343–349.

Orlóci, L. 1972a. On objective functions of phytosociological resemblance. Am. Midl. Nat. 88: 28–55.

Orlóci, L. 1972b. On information analysis in Phytosociology. In: E. Maarel (van der) & R. Tüxen (eds.), Basic Problems and Methods in Phytosociology, pp. 75–88. Junk, The Hague.

Orlóci, L. 1973. Ranking characters by a dispersion criterion. Nature 244: 371–373.

Orlóci, L. 1976. Ranking species by an information criterion. J. Ecol. 64: 417–419.

Orlóci, L. 1978a. Multivariate Analysis in Vegetation Research. 2nd edn. Junk, The Hague.

Orlóci, L. 1978b. Ranking species based on the components of equivocation information. Vegetatio 37: 123–125.

Patil, G.P., E.C. Pielou & W.E. Waters (eds.) 1971. Spatial Patterns and Statistical Distributions. Statistical Ecology, Vol. 1. The Pennsylvania State University Press, University Park.

Petraitis, P.S. 1979. Likelihood measures of niche breadth and overlap. Ecology 60: 703–710.

Pielou, E.C. 1969. An Introduction to Mathematical Ecology. Wiley-Interscience, New York.

Pielou, E.C. 1974. Population and Community Ecology: Principles and Methods. Gordon and Breach, New York.

Pielou, E.C. 1975. Ecological Diversity. Wiley, New York.

Podani, J. 1979. Association analysis based on the use of mutual information. Acta Bot. Acad. Sci. Hung. 25: 125–130.

Rajski, C. 1961. Entropy and metric spaces. In: C. Cherry (ed.), Information Theory, pp. 41–45. Butterworth, London.

Rényi, A. 1961. On measures of entropy and information. In: J. Neyman (ed.), Proceedings of the 4th Berkeley Symposium on Mathematical Statistics and Probability, pp. 547–561. Univ. of Calif. Press, Berkeley.

Rejmánek, M. 1981. Corrections to the indices of community dissimilarity based on species diversity measures. Oecologia 48: 230–231.

Sneath, P.H.A. & R.R. Sokal. 1973. Numerical Taxonomy. Freeman, San Francisco.

Shannon, C.E. 1948. A mathematical theory of communication. Bell System Tech. J. 27: 373–423.

Williams, C.B. 1964. Patterns in The Balance of Nature. Theoretical and Experimental Biology. Vol. III. Academic Press, London.

Williams, W.T. 1971. Principles of clustering. Ann. Rev. Ecol. Syst. 2: 303–326.

Williams, W.T. & J.M. Lambert. 1953. Multivariate methods in plant ecology. I. Association analysis in plant communities. J. Ecol. 47: 83–101.

Williams, W.T., J.M. Lambert & G.N. Lance. 1966. Multivariate methods in plant ecology. V. Similarity analysis and information analysis. J. Ecol. 54: 427–445.

Williams, W.T., G.M. Lance, L. Webb, J.G. Tracey & M.B. Dale. 1969. Studies in the numerical analysis of complex rainforest communities. III. The analysis of successional data. J. Ecol. 57: 515–535.

Yockey, H.P., R.L. Platzman & H. Quastler (eds.) 1958. Information Theory in Biology. Pergamon Press, New York.